Mathematical Recreations for the Programmable Calculator

Mathematical Recreations for the Programmable Calculator

Dean Hoffman
Auburn University

Lee Mohler
University of Alabama

HAYDEN BOOK COMPANY, INC.
Rochelle Park, New Jersey

To our parents, Gale and Vally Hoffman and James and Edith Mohler, and to our wives, Gail Coblick and Sherry Sullivan.

1	2	3	4	5	6	7	8	9	PRINTING
82	83	84	85	86	87	88	89	90	YEAR

Preface

In 1975, before the two authors met, we had both purchased HP-25 calculators and were immediately plunged into a mathematical fantasy land. Horrendous computations, slick mathematical models, bits of proofs of complex theorems, all of which were only imaginable before, could suddenly be realized on a little machine selling for an absurdly low price. After some time spent in something of a daze, the idea began to evolve that the amount of time spent with these calculators might be justified by writing a book about it. We spent the better part of the next four years producing this book. Although we share the same level of enthusiasm for programmable calculators, our interests and skills are complementary to the extent that what we have produced is certainly more than twice as good as what either of us could have done alone. While the concern of one was more with fairly elementary programming problems and with the actual organization and style of the book, the other was drawn to the more sophisticated and difficult problems. One wrote roughly the first half of each chapter and the other the second.

We would like to thank the staff of Hayden Book Company for their interest in and support for this project, especially Mr. Vernon Newton, who skillfully edited the entire manuscript. The father of one of us, James B. Mohler, with his talent and experience as a technical writer, made many helpful stylistic suggestions. We wish to thank Texas Instruments for giving us two TI-57s free of charge and Hewlett Packard for lending us an HP 33E without a charger. We thank Eileen Schauer for her elegant typing of the manuscript. Finally, and most especially, we thank Sherry Sullivan for painstakingly editing the entire manuscript before we submitted it to the publisher. Her efforts substantially clarified many portions of the text. Any errors that remain in the text are of course our responsibility; a book of this type is bound to contain some, and we would appreciate having them brought to our attention.

DEAN HOFFMAN
LEE MOHLER

Contents

Mathematical Recreations for the Programmable Calculator

*As the sun eclipses the stars by its brilliancy, so the man of knowl-
edge will eclipse the fame of others in assemblies of the people if
he proposes algebraic problems, and still more if he solves them.*
BRAHMAGUPTA (c. 600 AD)

Introduction

This book is a collection of recreational problems for programmable hand-held calculators. It was written with a twofold purpose. First, it is meant to be educational. The problems are designed
to help you discover some of the fundamentals of good programming
technique. Problems are to the programmer what exercises are to the
musician, a foundation for all future learning and development. If you
work all, or even half, the problems in this book, you will be well on
your way to mastering the art of calculator programming. In the process, you will also learn a lot of useful elementary mathematics.

Nevertheless, the book is not a textbook in programming. It
was also written for the purpose of recreation. A good deal of the
authors' "work" on the book consisted of countless hours spent simply playing with calculators. Programmable calculators are, after all,
marvelous toys—a fact not fully acknowledged by their manufacturers
or even by many professionals, but something most "ordinary" buyers
recognize immediately. As mathematicians, we have a ready appre-
ciation of the recreational potential of programmable calculators, and
a large part of what we are trying to do here is pass the fun along to
you.

You need only a limited amount of equipment and preparation
to enjoy this book. First, of course, you need a programmable calculator
and almost any one of them will do. Most problems in the book can be
solved on the smallest programmable calculators available, say a TI-
57 or an HP-25. The single crucial requirement is that your calculator
have test keys ($\boxed{x = t}$, $\boxed{x \geq t}$, $\boxed{x = y}$, $\boxed{x \geq y}$, and the like). You don't
need many, but you must have a few. This requirement obviously rules
out a few smaller calculators, such as the TI-55. A few of the problems
require larger calculators, and these are clearly marked. Owners of

1

large calculators, however, should not get the idea that these are the only problems meant for them. The main ingredient in the solution of all the problems is thought, and the thought required for any one problem changes very little when it is worked out on larger calculators.

Second, you need to understand the fundamentals of operating your calculator. Because this book is not based on any particular model of calculator, we must assume that you know how to carry out routine calculations on the model you own.[1]

As far as mathematical background goes, you need only a little algebra and, on rare occasions, some trigonometry, but we assume that you have some general interest in, and aptitude for, mathematics. Otherwise you wouldn't own a programmable calculator and you wouldn't have bought this book! On the other hand, you don't have to be a mathematical whiz to appreciate and enjoy the book. In fact, we have purposely written the problems with a wide range of difficulty, from several that are quite easy to a few that are very difficult.

The problems are grouped into three chapters. Chapter 1 is devoted to *technique*. Most of the problems develop methods that are useful for attacking other problems in the book. We have designed the book to allow the reader to enter it at any point and start reading. With this in mind, we have tried to make each problem as independent as possible of the others. Nevertheless, there are certain techniques—such as multiple storage—occurring often enough to justify special separate treatment. These are gathered together in Chap. 1.

Other problems in the first chapter—Loading Data, for example—are useful in a broader context; they are simply fundamental to all programming. Still others are more specialized and have been included because they are interesting in their own right.

Chapter 2 is devoted to numerical recreations. These can be described as a kind of intellectual sport man has engaged in for millennia, and to which the modern day calculator has added a new dimension. Many of the problems are taken from number theory, the oldest, most "useless," and at the same time most revered branch of pure mathematics.

Chapter 3 deals with games. There is a close connection between mathematics and games; indeed, some philosophers have taken the position that mathematics itself is nothing but an elaborate game. At any rate, it is true that many games admit a mathematical de-

---

[1] Let us take this opportunity to offer some advice: Read your manual! The more you know about the nuts and bolts of operating your calculator, the more skillful you will become at programming. The engineers who designed your calculator worked hard to pack as much as they could into a small space. You owe it to yourself to exploit fully what they have produced.

scription and consequently can be put onto a programmable calculator. In some of the problems in Chap. 3, the calculator functions as game equipment, while in others it takes on the more active role of player.

Following Chap. 3 are four short appendices and a glossary. The appendices are technical aids for using and understanding the rest of the book. They include (Appendix A) instructions on how to read flowcharts; (Appendix B) tips on troubleshooting (what to do when your program won't run); (Appendix C) a list of notations for particular calculator keys; and (Appendix D) actual program listings for a few programs in the book that fit very tightly onto small calculators.

The glossary is a collection of calculator and programming terms designed to familiarize you with the basic language of the subject. It contains short definitions of all the technical words used in the book. When you come across a word you don't know, therefore, look it up! Becoming conversant in the terminology of calculator programming will greatly increase your ability to learn your way around the subject, and it will give you a gratifying sense of professionalism as well.

The optimal way to read this book is with a calculator, pencil, and paper. The problems are all presented in short, straightforward sections laying out the necessary background, and frequently the discussion involves sample computations for the reader to do. Beyond, that you should have a pencil and calculator ready to do your own experimenting. (In fact, all mathematical books should be read this way because it is by *doing* mathematics that one achieves real understanding of it.) Some sections take you through a sequence of problems leading to one that would be too hard all by itself. Others contain more than one problem simply because the problems logically go together.

Since Chap. 1 is devoted to fundamentals, you may want to start there, especially if you don't have much experience with your calculator. In working through the book, however, we urge you to try to figure things out for yourself whenever possible. Each problem is presented with enough accompanying information to give you a reasonable chance of solving it. You should not look at our solution until you have tried the problem! We could almost say, do not look at the solution until you have *solved* the problem. If you cannot solve it without additional help, read the beginning of the solution and then return to the problem and try again.

Our solutions should be viewed only as suggestions. We have not always given the cleverest ones possible but have aimed instead for clarity. On the other hand, if a problem provides us an opportunity to use a valuable trick, we have done so. Now, it is axiomatic that any program can be improved in some way; there is no single right answer to a programming problem. You should therefore try to improve on our solutions. Incorporate whatever ideas you have had for solving a problem into our solution and make it your own. Ideally, of course, you will have produced your own solution without having to look at ours. In this case, you should read ours with the idea of picking up some new wrinkles you may not have thought of.

Many of the problems and solutions are followed by notes. Sometimes the notes comment or expand on the techniques used in the solution. Sometimes they relate the problem to others in the book or to mathematics or programming in general. Sometimes they place the problem in the more human perspective of the historical evolution of mathematics.

As we mentioned above, our solutions are not written around any particular calculator. Each consists of a careful statement of the techniques required and a description of the program steps in their proper sequence. Almost all solutions are accompanied by flowcharts. If a solution is very simple, there may be only a bare description of it along with the flowchart. All, in any case, are written with the presumption that you have worked on the problem and have a grasp of its main features.

We have assigned every problem in the book a difficulty rating, ranging from 1 (easy) to 4 (hard). We have also placed the easier problems toward the front of the chapters and have given their solutions in somewhat greater detail.

You no doubt know that there are two basic language types or "logics" for calculators, *algebraic* and *reverse Polish*. The most popular algebraic calculators are made by Texas Instruments, and the most popular reverse Polish by Hewlett-Packard. At times throughout the book, we comment on the slight differences in technique required by the two systems. We may use "TI" interchangeably with "algebraic" and "HP" interchangeably with "reverse Polish" in these discussions. You probably have a preference for the type of logic your own calculator uses, but you should not let this preference cause you to ignore the other language. For there will be times when you must understand program listings in the other language in order to adapt them to your own calculator. Some of the parallel discussions in this book will help familiarize you with that other language.

The fact that new programmable calculators are coming on the market all the time makes it difficult for a book like this to keep apace. However, we have been pleased to discover that these problems as they stand are perfectly suited to the newer models. We would change only two things if we were to begin writing the book again today. First, we would make more systematic use of subroutines, since virtually all programmable calculators now have them. Although we use subroutines quite a bit in the book, we do not assume that they are built in. Second, we would have something to say about alphanumerics. Programmable calculators with alphanumeric displays are now making their appearance, an example being the HP-41C. This new capability greatly improves the flexibility of a program's input and output. Nevertheless, calculators like the HP-41C, although they are very elegant, are still quite similar in basic design to their predecessors and work very well with the material in this book.

What is a trick the first time one meets it is a device the second time and a method the third time.

W. J. LEVEQUE
Fundamentals of Number Theory

CHAPTER ONE

Technique

1.1 Introduction

An artist friend of ours once said that in painting, "The greatest freedom comes from the greatest discipline." The same is true in programming. There are certain fundamental techniques the programmer needs to master so thoroughly that they need no further thought. Only then will he be free to tackle more subtle and difficult problems. They simply become the language he uses to express himself in a program. The more of these techniques you yourself learn and the more practiced you become in their use, the more freedom you will have in creating programs.

This chapter discusses all the techniques used elsewhere in the book. Nevertheless, you need not work your way completely through it before moving on. You may want to plunge directly into other chapters and return to this one when the need arises. If a problem requires a special technique, you will usually be directed to the appropriate section in this chapter. On the other hand, we think the problems posed in this chapter are interesting enough in their own right to be studied by themselves. If you master most of the techniques presented here, you will have acquired a solid foundation for more advanced programming.

This chapter falls roughly into two halves. Sections 1.2 through 1.9 involve the more fundamental material; Secs. 1.10 through 1.16 are more specialized.

1.2   Ten Quickies

Difficulty: 1 and 2

This section of ten topics—briefly handled and hence called "Quickies"—is designed to familiarize you with some little tricks of the trade that we have found useful. The problems are for the most part simple and straightforward once you have digested the preceding discussions. There are, however, a few cute ones.

In all the solutions, we use program listings rather than flow-charts. We do so because we are dealing here with fine details, with how to accomplish efficiently some of the things that flowcharts simply tell you to do without saying how. The programs are usually incomplete; for example, they frequently have no R/S order in them. The reason is that these routines are usually parts of larger programs.

As a rule, we give the listings in both algebraic and RPN languages. Read and compare both listings; the exercise will help you learn to translate from one language to the other. In two of the problems we give RPN owners the opportunity to do their own translations.

This section could have been extended almost indefinitely. If you have been paying close attention to your own programming, you probably have learned many useful little tricks that are not described here. It is good to think about these things as you work. They are the bread and butter of good programming.

(1)   Do Nothin' Till You Hear from Me: NOP

In a wide variety of mathematical systems, there is an object which plays the role of nothing. In arithmetic it is the number *zero*. In set theory, it is the *empty set*. In the language of programmable calculators, it is the *NOP* ("No Operation") *key*. When this step is encountered in a program, the calculator does nothing; it simply proceeds to the next step. At first, the whole business seems a little weird. Why would you ever want to use such a step in a program? The concept of "nothing" always has a paradoxical quality about it. By giving it a name, we turn it into "something," and yet what we are naming is the exact opposite of "something." No wonder the concept seems artificial.[1]

Well, God wouldn't have put that key on your calculator if he hadn't intended for you to use it. So let's take a closer look. Basically, NOP allows you to write a program that can be easily modified. If you think about it, you can see that you would not put NOP into a program unless you were going to replace it by something else at one time or

---

[1] As Alfred North Whitehead said of the number zero, "You don't go to the store to buy zero fish."

another. If you had intended to leave it there permanently, you could obviously have done just as well without it.

NOP can be used as a switch for turning certain parts of a program on and off. For example, let A represent a block of steps that is to be executed only at certain times. Put A at the bottom of the main program. At that point in the main program where A is to be executed, insert GTO A (or whatever GTO order is appropriate for your calculator). At the end of block A, insert another GTO order returning control to the step immediately following GTO A. If you want to turn off block A, just replace GTO A with NOP. (See Fig. 1-1.)

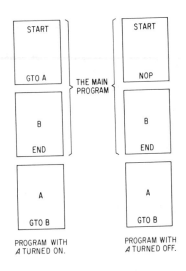

Fig. 1-1 NOP used as a switch

Sometimes NOP is used to turn off an individual step in a program, usually a PAUSE. PAUSEs at appropriate places allow you to watch a program in action. When you just want the program to run as fast as possible, the PAUSEs are replaced with NOPs. See Ulam's Problem (p. 251) for an example. NOP can even be used to turn off half of a key. To use [Dsz] solely for decrementing a memory register, for example, place a [NOP] immediately after it to prevent the program from branching.

If your calculator lacks [Ins] and [Del] keys, NOPs can be used in the developmental stages of a program. They permit extra steps to be inserted wherever they themselves are located, thus permitting the program to be altered without a wholesale reorganization (only short blocks will have to be moved). Moreover, deleted steps (including any PAUSE or R/S used to monitor the program) can easily be replaced by NOPs. A problem involving NOP will be discussed in Quickie No. 6.

(2) The ⌈+/−⌉ and ⌈1/x⌉ Keys

Two keys on the calculator that tend to be underused are the ⌈+/−⌉ key, which changes the sign of the number in the display (this key is labeled CHS on some calculators) and the ⌈1/x⌉ key, which takes the reciprocal of whatever is in the display. Perhaps their neglect is due to the fact that we can do without them. One can calculate 1/x by dividing 1 by x, and one can change the sign of x by subtracting it from 0. But since the ⌈+/−⌉ and ⌈1/x⌉ keys allow us to perform these operations in one step, we *should* use them for the sake of efficiency alone.

The real problem, however, is determining when to perform these operations in the first place. And this brings up a more likely reason why the keys are not used: the operations they represent rarely appear explicitly in algebraic formulas. One has to *invent* ways to use them in situations where they are not clearly called for. But it is worth learning such tricks, for the payoff is the saving of a program step or memory. If you have a small calculator, such a saving can make the difference between getting a program on your calculator or not.

Here is an example. Suppose that the number n is in your display, and you want to compute (1 − n). The n is in an awkward position. You would like to use the key sequence ⌈1⌉, ⌈−⌉, ⌈n⌉, ⌈=⌉ (*algebraic*) or ⌈1⌉, ⌈ENTER⌉, ⌈n⌉, ⌈−⌉ (*RPN*), but as soon as you push ⌈1⌉, you are going to lose n. One solution is to store n and bring it back when you need it: ⌈STO⌉ ⌈0⌉, ⌈1⌉, ⌈−⌉, ⌈RCL⌉ ⌈0⌉, ⌈=⌉ (*algebraic*) or ⌈STO⌉ ⌈0⌉, ⌈1⌉, ⌈RCL⌉ ⌈0⌉, ⌈−⌉ (*RPN*). A more efficient solution is to make use of the fact that changing sign and adding is the same as subtracting. Thus the following sequences will compute (1 − n), starting with n in the display: ⌈+/−⌉, ⌈+⌉, ⌈1⌉, ⌈=⌉ (*algebraic*) or ⌈+/−⌉, ⌈ENTER⌉, ⌈1⌉, ⌈+⌉ (*RPN*). Note that one step has been shaved off the algebraic sequence, and both sequences do away with the need for a memory. On some RPN calculators the ⌈ENTER⌉ in the above sequence is unnecessary, so a step can be taken off that sequence too.[2] Check this out on your own calculator.

Another way to use ⌈+/−⌉ in this problem is a consequence of the identity: (1 − n) = − (n − 1). This says that we can find (1 − n) by first finding (n − 1) and then changing its sign. Starting with n in the display, the sequence looks like this: ⌈−⌉, ⌈1⌉, ⌈=⌉, ⌈+/−⌉ (*algebraic*) or (⌈ENTER⌉), ⌈1⌉, ⌈−⌉, ⌈+/−⌉ (*RPN*).[3]

Now you may be asking yourself: What is n doing in the display in the first place? If I wanted to compute (1 − n), why didn't I

---

[2] The necessity of ⌈ENTER⌉ depends on where n came from and how your particular calculator behaves.

[3] Again, the ⌈ENTER⌉ may or may not be necessary.

first push $\boxed{1}$, $\boxed{-}$ or $\boxed{1}$, $\boxed{\text{ENTER}}$ and *then* put n in the display? If you were doing the computation by hand, of course, that's what you would do, but inside a program, things are not always so conveniently arranged. For instance, n might be the output of a subroutine, or it might be the input for the whole program, as in the following example.

Say you want to construct a program for plotting points on the graph of the equation, $y = \sqrt{9 - x^2}$.[4] The program will take as input the x-coordinate of a point on the graph and is to return the y-coordinate from the above formula. The user will thus put x in the display, and the program will compute $\sqrt{9 - x^2}$. The clever way to do this is to square x, change its sign, add the result to 9, and take the square root of the whole thing: $\boxed{x^2}$, $\boxed{+/-}$, $\boxed{+}$, $\boxed{9}$, $\boxed{=}$, $\boxed{\sqrt{x}}$ (*algebraic*) or $\boxed{x^2}$, $\boxed{+/-}$, $\boxed{9}$, $\boxed{+}$, $\boxed{\sqrt{x}}$ (*RPN*). It can also be done, as in the previous example, by using the identity: $9 - x^2 = -(x^2 - 9)$.

Similar tricks can be employed for performing divisions, starting with the divisor in the display.

Problem 1: Construct the most efficient possible key sequence for computing values of the expression, $1/(1 - x)$.

Problem 2: Construct a program for computing the intensity of illumination from a k-candlepower light source at various distances from the source. The constant k will be stored in memory m_0. The distance from the source will be the input for the program. If x is the distance (in feet) from the source, then the intensity of illumination is given by the formula, $I = k/x^2$.

(3) *Register Arithmetic*

Register arithmetic refers to operations performed in a memory rather than in the display. For example, the key sequence, $\boxed{\text{SUM}}$ $\boxed{0}$ (which is $\boxed{\text{STO}}$ $\boxed{+}$ $\boxed{0}$ on some calculators), adds the contents of the display to memory register m_0. The result is placed in m_0, while the display remains unchanged. Now suppose that we want to add the contents of m_0 and m_1. Ordinarily, we would use the following sequence:

Algebraic	RPN
RCL 0	RCL 0
+	RCL 1
RCL 1	+
=	

---

[4] The graph is a semicircle. Note that if we square both sides, the equation takes the form, $y^2 = 9 - x^2$, which can be rewritten as $x^2 + y^2 = 9$.

We could, however, use register arithmetic instead:

Algebraic	RPN
RCL 1	RCL 1
SUM 0	STO + 0

Note that the key sequences are shorter. The sum is also in m_0 rather than in the display. Of course, we could get it into the display by adding $\boxed{\text{RCL}}$ $\boxed{0}$, but sometimes it is better to have the sum in a memory so that the display can be used for other things, particularly in programs involving long sums (See Sec. 1.8).

Let us present another handy use of register arithmetic as a problem: It frequently happens that the output of a program winds up in some memory, say m_0, and at the end of the program you want to recall the contents of m_0 into the display and stop. At the same time you want to clear m_0 in preparation for the next run of the program.

Problem: Construct a key sequence for getting the contents of m_0 into the display and clearing m_0 without clearing the display.

(4) Parallel Computations

Another good use of register arithmetic arises when complicated mathematical expressions or more than one expression involving the same independent variable(s) need to be evaluated. The idea is to do part of the computation in the display and another part simultaneously in a memory. Here is a fairly simple example: Suppose that you want to write a program to compute values of the expression, $(x + y)/(x - y)$, where x and y are stored in memories m_0 and m_1, respectively. The straightforward way to make this computation is as follows:

Algebraic	Memories		RPN
RCL 0	0	x	RCL 0
+	1	y	RCL 1
RCL 1			+
=			RCL 0
÷			RCL 1
(			−
RCL 0			÷
−			
RCL 1			
)*			
=			

* The right parenthesis can probably be deleted from this program (see Quickie No. 10).

All computations were performed in the display. A more efficient method is to compute (x − y) in memory m_0 while (x + y) is being computed in the display, as follows:

Algebraic		Memories		RPN
RCL 0		0	x	RCL 0
+		1	y	RCL 1
RCL 1				STO − 0
INV SUM 0				+
=		0	x − y	RCL 0
÷		1	y	÷
RCL 0				
=				

Try this sequence out by hand on your calculator. As you execute each step, stop and picture what is happening. Learning this trick will save steps in your programs time and again.

The next example saves only one step in a long sequence, but it gives us an opportunity to introduce (or hopefully recall) a formula every educated person should know: the law of cosines.

The law of cosines states that given any triangle ABC with sides a, b, and c and angles α, β, and γ (see Fig. 1-2), the following equation holds:

$$c^2 = a^2 + b^2 - 2ab \cos(\gamma)$$

This formula expresses the length of side c of the triangle in terms of the lengths of the other two sides and the angle between them. Taking square roots on both sides, we obtain:

$$c = \sqrt{a^2 + b^2 - 2ab \cos(\gamma)}$$

An observer situated at point C can use this formula to determine the distance between two remote points A and B if the distances between himself and A and B are known, as well as the angle between the two lines joining his position to points A and B. This technique is particularly useful if the triangle is a large one, say a triangle formed by an observer in a control tower and two airplanes in flight.

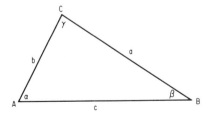

Fig. 1-2 An ordinary triangle

Problem: Construct a program for computing c from the above formula given a, b, and γ. Assume that these are placed in memories m_0, m_1, and m_2, respectively.

(5) The $\boxed{\text{Dsz}}$ Key

Before reading this section, you need to know something about loops. If you don't know what a loop is or how one works, see Sec. 1.8.

The $\boxed{\text{Dsz}}$ key is a one-step operation used for controlling loops ("Dsz" stands for "Decrement, skip if zero"). Typically, a loop is controlled by some kind of counter, which, in its simplest form, works like this: Some memory, say m_0, is used to store the number of times, n, that we want the loop to be executed. Each time through the loop, a 1 is subtracted from the contents of m_0, and a check is run to see if $m_0 = 0$. If so, the loop has been executed n times and is terminated. If not, control is sent back to the top of the loop. The flowchart is shown in Fig. 1-3.

When coupled with the step following it in a program, $\boxed{\text{Dsz}}$ executes this flowchart. When $\boxed{\text{Dsz}}$ is executed on the TI-57, for instance, it subtracts 1 from the contents of m_0 and then checks to see if $m_0 = 0$. If not, control moves to the following step in the program, a GTO order (perhaps RST) that returns control to the top of the loop. If $m_0 =$ zero, the following step is skipped, and the program continues from the next step.

On other calculators, the key works essentially the same way, but with minor variations. In HP calculators, the $\boxed{\text{Dsz}}$ button decrements a special index register (I) instead of m_0. In some fancier TI calculators, the $\boxed{\text{Dsz}}$ step can be performed on more than one memory, the choice being indicated by a number following the $\boxed{\text{Dsz}}$. Many loops can be efficiently controlled in this way. We will now present a problem with a very short solution.

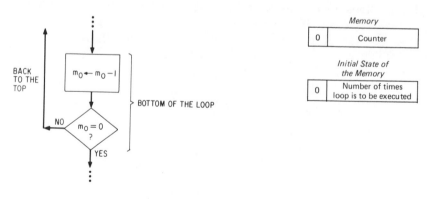

Fig. 1-3 Flowchart for the bottom of a loop

Problem: Write a program that sums up all positive whole numbers from 1 to n (n is to be put in by the user):

$$1 + 2 + 3 + \cdots + n$$

(6) Alternate Uses of the Skipped Step

All calculator testing functions make use of the device of skipping a program step if the answer to a test question is "no," the skipped step usually being a GTO order that sends control elsewhere. This device allows a calculator (or computer) to make "decisions." Two numbers are compared in some way by the test question. Depending on the outcome of the comparison, the program executes one sequence of steps or another—those accessed via the GTO order or those immediately following the (skipped) GTO order. On some calculators, for example, the TI-58/59, the step in a program immediately following a test question *must* be a GTO order, but on others, like the TI-57, HP-25 and HP-67/97, this step can be anything. Such flexibility can be put to some good uses.

Here is an example: Suppose that you have a pair of numbers, n_1 and n_2, and that you want to get your hands on the larger of the two (see Sec. 2.15, for example). Put one of the numbers, say n_1, in the x-register and the other, n_2, in the y-register (t-register on some calculators). Now you want to get the larger number in the x-register and the smaller in the y-register. If the larger number just happens to be in the x-register, you don't want to do anything with it at the present time; you just want to go on with the program. If the larger number is in the y-register, you want to transpose x and y and then proceed with the program. You can do this by inserting the single step $\boxed{x \rightleftarrows y}$ (or $\boxed{x \rightleftarrows t}$) immediately following the test, $\boxed{x < y}$ ($\boxed{\text{INV}}$ $\boxed{x \geq t}$ on some calculators). If the contents of the x-register is smaller, the $\boxed{x \rightleftarrows y}$ will get executed; if not, the step will be skipped, which is exactly what you want to happen.

Problem 1: Construct a program for computing the absolute value of a number. (Since calculators already have a key for this function, the idea, of course, is *not* to use that key.)

Problem 2: Devise a switch for turning off a test so that—when the switch is set—the program will behave as if the test were not there.

(7) Large Constants

There is a rough tradeoff in programming between the number of steps and memories used. Steps can usually be eliminated from programs by increasing the use of memories, and, conversely, the number of memories can be decreased by lengthening the program.[5]

One should be aware of this tradeoff and try to exploit whichever mode will produce the most efficient overall program. For example, it is definitely more advantageous to use memory for programs that return values of a formula with large constants. Suppose, for example, that we want to write a program for converting light years to inches. Since there are approximately 3.72×10^{17} inches in a light year, the program will multiply the input (distance in light years) by 3.72×10^{17} to produce the desired output. Keying the constant 3.72×10^{17} directly into the program where needed will cost seven program steps— 3 · 7 2 EE 1 7 —since the program memory stores only one keystroke at a time. The program will therefore look like this (assuming that the number to be converted is sitting in the display when the program starts):

Algebraic	RPN
x	Enter
3	3
.	.
7	7
2	2
EE	EEX
1	1
7	7
=	x

It would be much more efficient to store the constant in memory and recall it when needed, using only a single step. The program will now look like this:

Algebraic	Memories		RPN
x	0	3.72×10^{17}	RCL 0
RCL 0			x
=			

For this particular problem, it doesn't make a lot of difference which program is chosen since your calculator clearly has room for either, but in a large program, the saving of six steps is a significant matter. When a formula to be programmed has to be embedded in a larger program, for example, the amount of space available for it may

---

[5] Attention, TI-58/59 and HP 41C owners. This is a general remark. It does not refer to the feature of your calculator that allows you to trade program space for addressable memories. However, the fact that your calculator permits the trade is another indication that the general remark is true.

be quite limited. Section 1.4 contains a program of this type. There we evolve a skeleton program for finding the maximum output that can be generated by a given formula with one independent variable. A program for this formula,

$$I(x) = (1 + .05x)(50000 - 5000x)$$

has to be embedded in the skeleton program. At this time, your problem is to write as short a program as you can for computing values of $I(x)$.

(8) Large Numbers

A calculator can handle numbers so large you would think overflow would never be a problem. Yet it is easy to concoct numbers too big for the calculator to handle. Try calculating 5^{250}, for example. Should such a number occur as program output, there is not much you can do about it. It is just too big for the calculator. Sometimes, however, large numbers occur as intermediate results in a computation whose final answer *can* be handled by the calculator. The problem then is to coax the calculator past these difficult places in the computation. The solution is sometimes just a matter of controlling the order in which operations are performed.

Suppose, for example, that you want to compute the following number:

$$\frac{(3.14 \times 10^{58})(2.8 \times 10^{60})}{(5.28 \times 10^{16})(9.7 \times 10^{90})}$$

If you try to compute the product in the numerator or denominator first, you will get into trouble. The result will be much too large. The trick is to alternate multiplication and division, but even here you have to be a little careful. If you begin by dividing (3.14×10^{58}) by (5.28×10^{16}) and then multiplying by (2.8×10^{60}), the result will still be too big (a calculator cannot handle any exponent larger than 99). You should therefore first divide (3.14×10^{58}) by (9.7×10^{90}), then multiply by (2.8×10^{60}), and finally divide by (5.28×10^{16}), although other orders will work too.

There is a difficulty with such computations. You have to examine the numbers involved before deciding on the order of operations—not simple to do if the numbers have been generated *inside* a program. Fortunately, there is a trick for dealing with very large numbers on a calculator, a trick that you should already have some familiarity with: logarithms. We will base our discussion on the natural log function, which appears as $\boxed{\ln x}$ on your calculator. Your calculator also has another log button, $\boxed{\log x}$, the common logarithm. Its properties are essentially the same as $\boxed{\ln x}$, only on a different scale.

The value of logarithms is that they reduce numbers to a much smaller scale. If you take the logarithms of a group of numbers, the logs keep the same order as the numbers they came from but are much much closer in value to one another. Compute the natural logs of 1, 2, 3, 4, 5 on your calculator, for example. The results run between 0 and 1.60943791—a rough spread of 1.6 rather than 4. Compression gets greater as the numbers get larger. The natural logs of 100, 200, 300, 400, and 500 are compressed between about 4.605 and 6.215.[6] Now try taking the natural log of the biggest number a calculator permits: $9.999999999 \times 10^{99}$ (your particular calculator may accept a few digits more or less). The result is about 230. Then try the smallest positive number on your calculator: 1×10^{-99}. The result is about −228. Thus the entire range of positive numbers on your calculator gets compressed into the comparatively tiny range between −228 and 230 by the natural log function.[7] Obviously, plenty of room is available for the logarithms of numbers that are too big (or too small) to fit on the calculator themselves.

The scheme for dealing with numbers so large that intermediate results may run off the calculator, therefore, is to compress them by taking their logarithms, to do the necessary arithmetic in their compressed scale, and then to re-expand them for the final result. Because of the nonuniform nature of the compression, noted above, arithmetic in the compressed scale will not be the same as in the ordinary scale. But the marvelous fact is it actually gets simpler! Multiplication and division become addition and subtraction in the compressed scale, and exponentiation (y^x) and root extraction (INV y^x or $\sqrt[x]{y}$) become multiplication and division. Although there are no analogs in the compressed scale of addition and subtraction in the real-world scale, these two elementary operations lead to overflow problems much less often than the other four.

The precise formulation of the connection between ordinary arithmetic and logarithmically compressed arithmetic is contained in the four rules of logarithms:

(1) $$\ln(ab) = \ln(a) + \ln(b)$$

(2) $$\ln(a/b) = \ln(a) - \ln(b)$$

(3) $$\ln(a^n) = n[\ln(a)]$$

(4) $$\ln(\sqrt[n]{a}) = \ln(a)/n$$

Rule 1 says that to find the natural log of the product of two numbers (that is, to find where the product of two numbers lies in the

---

[6] Note that the difference between ln(5) and ln(1) is exactly the same as the difference between ln(500) and ln(100). This is not a coincidence.

[7] Logarithms, however, are not defined for negative numbers.

compressed scale), you add the natural logs of the two numbers. Rule 2 says that to find the quotient of two numbers in the compressed scale (that is, to find the natural log of the quotient of two numbers), you subtract their natural logs. Thus, to find the natural log of a/bc, for example, you would take the natural log of a and subtract the natural logs of b and c. Or you could add the natural logs of b and c and subtract the result from the natural log of a. To put this symbolically,

$$\ln\left(\frac{a}{bc}\right) = \ln(a) - \ln(b) - \ln(c) = \ln(a) - [\ln(b) + \ln(c)]$$

We will leave you to sort out the workings of rules 3 and 4.

Now that you know how to do multiplication and division in the compressed scale, there is only one problem left: How do you get back to the scale you started from? It's easy. You just apply the inverse natural log function, which turns out to be e^x (it is called $\boxed{\text{INV}}$ $\boxed{\ln x}$ on some calculators).[8]

Here is a complete prescription for finding the product of two numbers, a and b: First, take the natural logs of a and b and add them together, giving you ln(ab). Then recover ab by taking $\boxed{e^x}$ of this result ($\boxed{\text{INV}}$ $\boxed{\ln x}$ on some calculators). Following are two typical key sequence listings for hand calculations.

TI-57	HP-25
(key in a)	(key in a)
ln x	ln x
+	(key in b)
(key in b)	ln x
ln x	+
=	e^x
INV ln x	

Try this out on a couple of numbers for which you already know the answer. You will see that it works.[9]

Now it is time for you to start doing these tricks for yourself. Here are two problems:

1. Write a program for computing ab/cd using logarithms.
2. Write a program for computing a^b.

(9) Accuracy

We are going to discuss two kinds of accuracy here: (1) the problem of getting the best possible result in a chain calculation (one

---

[8] The relevant identity here is: $e^{\ln x} = x$.

[9] If your answer is slightly off, see the notes for an explanation.

involving several operations), and (2) the problem of obtaining absolutely precise results when dealing with whole numbers.

Let's begin with a brief tour of all the numbers your calculator is capable of expressing, assuming, for the purpose of this discussion, that your calculator has a ten-digit display. Since a number is just a string of digits (let us ignore signs, decimal points, and exponents for the moment), any string from 0 up to 9999999999 can be expressed with a ten-digit display. That is 1×10^{10} numbers already. Now, how many numbers can you express in scientific notation? Inserting a decimal point between the first and second digits of the ten-digit string and following the string with a two-digit exponent provides a typical number in scientific notation. As we just noted, there are 1×10^{10} possibilities for the mantissa (the ten-digit string with a decimal point inserted), and the exponent can be anything from -99 to 99, or 199 possibilities. Consequently, there are $1 \times 10^{10} \times 199$, or 1.99×10^{12} possible *positive* numbers in scientific notation. When the negative numbers are added, the figure doubles, to 3.98×10^{12}. This, in fact, is everything. Since any number you put into your calculator *can* be put in scientific notation (your calculator may not be able to show you the entire mantissa, but it's still there, inside), once you've counted all the numbers possible in scientific notation, you've got them all.

Rest assured that 3.98×10^{12} is a lot of numbers. If you devoted yourself 24 hours a day to nothing but looking at them one after another, and if you gave each one-tenth of a second of your time, it would take over 12,000 years to see them all! Even though your calculator is ready to show you any one of those 3.98×10^{12} numbers at any time, you will never actually see more than a tiny fraction of them.

Next, let's take a look at how these numbers are distributed. The distribution is far from uniform. The two smallest positive numbers on your calculator are 1×10^{-99} and 2×10^{-99}. The difference between them is 1×10^{-99}. Near zero, therefore, the numbers are very close together. How about near 1? The next number after 1 on your calculator is 1.000000001. The difference between these two numbers is 1×10^{-9}—still quite small, but 10^{90} times as large as the difference between 1×10^{-99} and 2×10^{-99}. And if you thought 3.98×10^{12} was big, you can imagine how large 10^{90} is.

The other extreme is reached in the difference between the two largest numbers on the calculator, $9.999999998 \times 10^{99}$ and $9.999999999 \times 10^{99}$, which is 1×10^{90}. Thus, vastly more whole numbers are *missing* between the largest two numbers on your calculator than are *present* in the calculator's entire domain of numbers. Another point worth noting is that there are exactly as many numbers in your calculator between 0 and 1 as there are between 10 and infinity. The former numbers are expressible as a 10-digit mantissa with an exponent between -99 and -1; the latter, as a 10-digit mantissa with an exponent between 1 and 99.

The image we are trying to build up in your mind is the following. Near zero, the numbers are very closely packed—inconceivably close. As you move away from zero, they become less and less dense. At the outer limits, the distance between successive numbers is inconceivably large. The moral of this story, as far as accuracy is concerned, is that your calculator is a lot more accurate near zero than it is away from zero, because the gradations are finer there. Thus, if you keep the intermediate results in a chain calculation as small as possible, you will achieve the best accuracy. This is only a general rule, however; the accuracy of the outcome of many types of computations depends only on the number of digits of accuracy in the mantissa. In these cases, it doesn't matter how big or small the numbers themselves are.

Now let's look at the other type of accuracy. There is a class of calculations that stays entirely within the domain of whole numbers, calculations that have to do, in general, with *counting*. The branch of mathematics devoted to this art of counting is called *combinatorics*. When doing combinatorial computations, we must be careful to avoid slipping away from whole numbers. Many combinatorial algorithms involve comparisons to see if two numbers, one or both of which are the product of some computation, are equal. If one or the other is ever so slightly off, we can get a "no" answer to an equality test when the answer should be "yes." Consequently, one must be sure all intermediate results in a combinatorial computation remain whole numbers. Note that this advice runs counter to the "stay close to 0" advice in the previous paragraph.

Here is a problem to solve involving accuracy. As we noted in the previous article, the y^x key does not produce accurate results for large numbers. Therefore, construct an accurate program for computing 2^n where n is a positive whole number.

(10)   The Stack

If you own a reverse Polish calculator, you know what a stack is. If your calculator is algebraic, it has a stack too, and you need to find out about it. The stack is where numbers are loaded in preparation for performing operations on them. In RPN calculators, the loading is done manually, using the [ENTER] and [RCL] keys. The operations are performed after loading by manipulating the stack with the [ENTER], [R↓], and [x⇌y] keys to move its contents into the proper locations.

In algebraic calculators, the stack is internal and not directly accessible to the user. While numbers and operations are being keyed into the calculator together, the calculator automatically loads the internal stack, storing the numbers *and* special codes for the operations. Because a *heirarchy* dictates that some operations must be executed before others even if they are keyed in later, the calculator monitors each operation as it is keyed. Sometimes it will execute an operation before the ⌐=⌐ key is pushed in order to save stack space. After the user has entered all numbers and operations, he presses the ⌐=⌐ key. The calculator then reads through the stack and performs the operations in the order dictated by the algebraic heirarchy.

The internal stack and the heirarchies that manipulate its contents are, in effect, extra memory space in your calculator. You need to learn how they work—that is, what the calculator is going to do each time you push a key—so that you can exploit them. Since you can't see inside the stack, you will not be able to figure a lot of things out (unless you get fanatical), but at least you can watch what the calculator does *externally* as a clue to what is happening inside. (RPN owners, don't go away! We'll have a problem for you a little later.)

First, let's make some observations about algebraic calculators. When you press the key sequence ⌐5⌐ and ⌐+⌐, the 5 and the + go into the stack. Notice that the 5 is now held in two places—in the stack as well as in the display. Sometimes you can exploit this fact. If you want to find $5 + \sqrt{5}$, for example, merely hit ⌐√x⌐ and ⌐=⌐. The key sequence for a program for computing $x + \sqrt{x}$, assuming that the user has already placed x in the display, could be written as follows: ⌐+⌐, ⌐√x⌐, ⌐=⌐. Here is another example. Since your calculator probably fills in any missing closed parentheses automatically when the ⌐=⌐ is pushed, leave them out if they occur at the end; the calculator will supply them. To find $(a + b)/(c + d)$, for example, just press ⌐a⌐, ⌐+⌐, ⌐b⌐, ⌐=⌐ ⌐(⌐, ⌐c⌐, ⌐+⌐, ⌐d⌐, ⌐=⌐.[10]

Another thing you need to know is that your calculator will execute some operations before the ⌐=⌐ is pressed if it decides that they will come first in the execution heirarchy no matter what you key in later. If you press ⌐3⌐ ⌐x⌐ ⌐5⌐ ⌐+⌐, for example, the calculator will carry out the multiplication, because once the ⌐+⌐ is pushed it knows that 3×5 must get executed first. This characteristic saves space in the stack. It also places the number 15 in the stack and in the display. Can you use it? Maybe you can. Try the following: Write a program for computing values of $ab + \sqrt{ab + c}$ when a, b, and c are placed in memories m_0, m_1, and m_2 by the user.

Now a word for RPN owners. You too can use the x-in-two places trick. When you press ⌐ENTER⌐, the number in the display moves up to the y-register, but it also stays in the x-register. If you act fast, you can use it. To illustrate, a program for $x + \sqrt{x}$ would be as follows: ⌐ENTER⌐, ⌐√x⌐, ⌐+⌐. But you really need to learn how best to exploit the

---

[10] Note that we used ⌐=⌐ on the (a + b) rather than parenthesis. It saves a step.

stack and its ordinary behavior by using it whenever possible. Suppose that you make some computation and will need the result later. It may be possible to leave it in the stack while you do other things, instead of storing it and having to recall it.

Here is another exercise in stack manipulation: Load your stack so that it looks like Fig. 1-4(a). Now figure out a key sequence for rearranging it into the pattern shown in Fig. 1-4(b).

t	4
z	3
y	2
x	1

(a)

t	1
z	2
y	3
x	4

(b)

Fig. 1-4 Stack manipulation

Solutions for the Quickie Problems

Quickie 2, Problem 1: The key sequences for computing $1/(1 - x)$ are as follows:

Algebraic	RPN
+/−	CHS
+	(ENTER)
1	1
=	+
1/x	1/x
or	or
−	(ENTER)
1	1
=	−
+/−	CHS
1/x	1/x

Quickie 2, Problem 2: For computing k/x^2 with k in m_0 and x in the display, we use the following:

Algebraic	RPN
x^2	x^2
1/x	1/x
×	RCL 0
RCL 0	x
=	
or	or
x^2	x^2
÷	RCL 0
RCL 0	÷
=	1/x
1/x	

Notes: On RPN calculators these problems can also be handled efficiently using the $\boxed{x \rightleftharpoons y}$ button. For instance, k/x^2 can be computed using the sequence $\boxed{x^2}$, $\boxed{\text{RCL}}$ $\boxed{0}$, $\boxed{x \rightleftharpoons y}$, and $\boxed{\div}$. Buttons $\boxed{+/-}$ and $\boxed{1/x}$ both have the interesting feature that if you push them twice, you're back where you started.[11] The only other button on the calculator with this property is $\boxed{x \rightleftharpoons y}$ or $\boxed{x \rightleftharpoons t}$.

Quickie 3: The solution is surprisingly simple. Recall the contents of m_0, and then use register arithmetic to subtract the contents of the display (which are now the same as those of m_0) from m_0. This maneuver turns the contents of m_0 into 0 and leaves the display undisturbed. The sequence on two popular calculators is as follows:

TI-57	HP-25
RCL 0	RCL 0
INV SUM 0	STO – 0

Quickie 4: We want to compute $\sqrt{a^2 + b^2 - 2ab\cos\gamma}$. We will use register arithmetic to compute the product ab while $a^2 + b^2$ is being computed in the display. When we compute $2ab\cos\gamma$ in the display, ab will be recalled. Here are the key sequences:

Algebraic	Memories		RPN
RCL 0	0	a	RCL 0
x²	1	b	x²
+	2	γ	RCL 1
RCL 1			STO × 0
Prd 0			x²
x²			+
–			2
2			RCL 0
×			×
RCL 0			RCL 2
×			cos
RCL 2			×
cos			–
=			√x
√x			

Quickie 5: The details of this solution apply to the TI-57, but the general outline will work on your calculator. Don't skip over the TI-57 program below. Translate it into your own calculator's language; it's good practice. We will form the sum in memory register m_1, stor-

---

[11] On some calculators, this doesn't quite work with 1/x, although in theory it should. Check it out on your own calculator.

ing n in m_0, the loop counter. The trick is to note that the summands we want will be generated in m_0 by the execution of the $\boxed{\text{Dsz}}$ order. Of course, they will appear in reverse order (n, n−1, n−2, $\cdots$, 1), but that won't make any difference. All we have to do inside the loop is recall the contents of m_0 and add it to m_1, using register arithmetic (see Solution 3). When the loop is terminated by a "skip zero," the sum we want will be in m_1. Here is the program listing for the TI-57:

Memories	
0	counter
1	sum

00	RCL 0
01	SUM 1
02	Dsz
03	RST (sends control back to 00)
04	R/S

Initial State of the memories

0	n
1	0

Notes: We have kept the program short for dramatic effect. It will be easier to use if we end with the contents of m_1 in the display and clear m_1 to be ready for the next run. See Solution 3 for a clever way to do this and Sec. 1.16 for a much more efficient algorithm for computing this same sum.

Quickie 6: To take the absolute value of a number, you leave it alone if it is positive and change its sign if it is negative. All you have to do is execute the test $\boxed{\text{x} < 0}$ [if your calculator doesn't have this test, you will have to put a 0 in the y-register (or t-register) and then run the test x < y (or x < t)] followed by the step $\boxed{\text{CHS}}$ ($\boxed{+/-}$). If the contents of the display is positive, the latter step will get skipped. Here are the sequences for the TI-57 and HP-25:

TI-57	*HP-25*
c.t.	x < 0
(clear t-register)	CHS
INV x $\geq$ t	
+/−	

To turn a test off, simply insert $\boxed{\text{NOP}}$ immediately after the test. Either the step will get skipped or it won't; in either case, it won't make any difference because the step doesn't do anything.

Quickie 7: The straightforward solution is to store the three constants 0.05, 50000, and 5000, in memories m_0, m_1, and m_2 and then compute I(x) as indicated by the formula, recalling the constants as they are needed. Because x appears twice in the formula, it too will have to be stored in the first step of the program. And since it will be sitting in the display at the beginning, there's no sense in letting it go to waste after the ⌐STO⌐ step; start multiplying by 0.05, therefore, on the way to computing $1 + .05x$. Here are the program listings:

Algebraic	Memories		RPN
STO 3	0	.05	STO 3
×	1	5000	RCL 0
RCL 0	2	50000	×
+	3	x	1
1			+
=			RCL 1
×	*Initial State*		RCL 2
(	*of the Memories*		RCL 3
RCL 1	0	.05	×
−	1	50000	−
RCL 2	2	5000	×
×	3		
RCL 3			
)			
=*			

* The second parenthesis is probably unnecessary (see Quickie No. 10).

Both of these programs can be shortened by rewriting the formula for I(x). Carrying out the multiplication indicated in the original formula and collecting the middle terms, we get: $I(x) = 50000 - 2500x - 250x^2$. This formula can be most efficiently evaluated by storing the three constants 50000, 2500, and −250. We store −250 rather than 250 to avoid having to change signs when $-250x^2$ is calculated in the beginning of the program. We could also store −2500 instead of 2500, but it turns out to make no difference. Once again x will have to be stored at the beginning, since it appears twice. We will evaluate the formula starting from the right so as to use the x already in the display. Here are the listings:

Algebraic	Memories		RPN
STO 3	0	−250	STO 3
x^2	1	2500	x^2
×	2	50000	RCL 0
RCL 0	3	x	×
−			RCL 1
RCL 1	*Initial State of the Memories*		RCL 3
×	0	−250	×
RCL 3	1	2500	−
+	2	50000	RCL 2
RCL 2	3		+
=			

Notes: Here is an RPN program for computing I(x) which is one step shorter than the one just given. We leave you to unravel the mystery of why it works.

	Memories
STO 3	
RCL 0	(Same as above)
×	
RCL 1	
−	
RCL 3	
×	
RCL 2	
+	

Quickie 8: For Problem 1, we will assume that a, b, c, and d are in memories m_0 through m_3. To find the natural log of ab/cd, add the natural logs of a and b and then subtract the natural logs of c and d from the result. Finally apply $\boxed{e^x}$ to recover ab/cd itself. Here are the program listings:

Algebraic	*Memories*		*RPN*
RCL 0	0	a	RCL 0
ln x	1	b	ln x
+	2	c	RCL 1
RCL 1	3	d	ln x
ln x			+
−			RCL 2
RCL 2			ln x
ln x			−
−			RCL 3
RCL 3			ln x
ln x			−
=			e^x
INV ln x			

Here are the program listings for Problem 2 [computations are dictated by logarithmic rule (3)]:

Algebraic	*Memories*		*RPN*
RCL 0	0	a	RCL 0
ln x	1	b	ln x
×			RCL 1
RCL 1			×
=			e^x
INV ln x			

Notes: In the precalculator era, logarithms were used much more widely than today. The simplifications of arithmetic produced by logarithms (invented in the seventeenth century by John Napier) made them virtually indispensible whenever computations involving unwieldy numbers had to be performed. Voluminous tables of logarithms, some accurate to as many as 20 decimal places, were published for this purpose—an enormous project involving hundreds of people and years of hand calculations on mechanical adding machines. Such calculations are now performed inside your calculator every time you press the lnx key, the calculations for an individual number taking less than a second. Logarithmic principles also formed the basis for the workings of the standard precalculator, scientific calculating device—the slide rule.

Calculators use logarithms in some of their internal computations to save you the trouble of having to do so. If you compute 2^{26} first using the program given in the solution to Problem 2 and then using your y^x key, you will probably come up with the same answer for both (the HP-67 is an exception). This shows that the calculator is probably using the program given above, because the answer happens to be wrong. It is clear that 2^{26} should be an even whole number. If you still think your calculator is producing the right answer, make another check. Take the fractional part of the answer (FRAC or INV INT). Since the correct answer must be a whole number, you should clearly get 0. Calculators sometimes hide little discrepancies in their guard digits.

The reason that computing 2^{26} using logs produces errors is that 2^{26} lies at the end of the scale, and here logarithmic compression is extreme. The loss of resolution in the compressed scale leads to errors. For a discussion of this issue with respect to the y^x key, see Sec. 1.10 and also Solution 9, below.

Quickie 9: The idea is to go back to the *definition* of 2^n: 2 multiplied by itself n times. We will use a loop and the [Dsz] key (see Solution 5). Start with n in the register that the [Dsz] key decrements. (If your calculator doesn't have a [Dsz] key, put n in m_0 and see Fig. 1-3 for a flowchart showing how to simulate this key.) Register m_1 will be used to build up 2^n through successive multiplications by 2. When the program starts, m_1 should have a 1 in it. Each time through the loop we will multiply the contents of m_1 by 2, using register arithmetic (see Solution 3). When the [Dsz] key has gotten the contents of the memory that held n down to 0, the contents of m_1 will have been multiplied by 2 for n times, and it will thus be time to quit. The desired output of the program will be found in m_1. The listing for this program on a TI-57, which is short enough for you to translate it onto your own calculator, is as follows:

		Memories		Initial State of Memories	
00	2				
01	Prd 0 (STO × 0)	0	counter	0	n
02	Dsz	1	answer accumulator	1	1
03	RST (sends control back to the top)				
04	R/S				

Notes: The program can be improved by adding steps before R/S that recall the contents of m_1 into the display and put a 1 in m_1, making it ready for the next run. Just two steps will be needed if we use a trick analogous to that in Solution 3. We leave it to you to figure out. (Since this program is also comparatively slow, see Sec. 1–10 for ·a faster but rather elaborate routine for solving the same problem and for a thorough discussion of the problem of computing y^x accurately when x and y are whole numbers.)

Quickie 10: To compute $ab + \sqrt{ab + c}$, start with $\boxed{a}$ $\boxed{\times}$ $\boxed{b}$ $\boxed{+}$. While the ab is still in the display, add it to c using memory arithmetic. Then recall this quantity and complete the calculation. Here is the key sequence for a TI-57:

RCL 0
×
RCL 1
+
SUM 2
RCL 2
$\sqrt{x}$
=

Memories	
0	a
1	b
2	c

To rearrange the stack, use the following sequence: R↓, x ⇌ y, R↓, x ⇌ y, R↓, R↓, R↓, x ⇌ y. Here is a picture of what happens to the stack:

4		1		1		3		3		4		2		1		1
3	R↓	4	x⇌y	4	R↓	1	x⇌y	1	R↓	3	R↓	4	R↓	2	x⇌y	2
2	→	3	→	2	→	4	→	2	→	1	→	3	→	4	→	3
1		2		3		2		4		2		1		3		4

If your stack will roll in both directions (that is, if you have R↑), the following shorter sequence will work: R↓, x ⇌ y, R↓, x ⇌ y, R↑, x ⇌ y.

Notes: Looking at the stack reveals the difference between algebraic and RPN calculators. Algebraic calculators are more automated. Since they incorporate a hierarchy taken over from written algebraic expressions, such expressions can be translated directly onto the calculator more easily. Although RPN calculators make the user unravel the hierarchy and decide in what order things should be done, they allow him more freedom and flexibility by giving him direct access to the stack. You have probably noticed that RPN programs usually turn out to be the shorter of the two. The difference between the calculators is something like that between automatic and manual transmissions: one is easier to use but the other gets better gas mileage.

1.3   Test Functions

Difficulty: 1

Calculator programs automatically carry out procedures that would otherwise have to be done by hand. When we do calculations by hand on a calculator, we engage in two different types of activity: (1) We press the keys, and (2) we interpret the results that appear in the display, the interpretation often requiring us to make a *decision* about which keys to press next.

Let's look at an example. Suppose that you are going to put some money into a savings account paying interest at an annual rate of 8 percent, compounded quarterly. At the end of each compounding period (3 months), your money will thus be multiplied by 1.02 (that is, it will increase by 2 percent). How long, then, will it take for your money to double? Well, at the end of n compounding periods, your money will have been multiplied by 1.02 n times, that is, it will have been multiplied by 1.02^n. Since your money will be doubled when it has been multiplied by 2, the question is: When does $1.02^n = 2$? You could solve the equation, $1.02^n = 2$, for n, but suppose you have forgotten how? Then you will get out your calculator and simply compute 1.02^n for various values of n until you find the n such that 1.02^n equals 2. As each value of 1.02^n is calculated, you will compare it with the value you are trying to get, 2. If it is too large, you will adjust n downward. If it is too small, you will increase n. Eventually, you will zero in on the correct value, which turns out to be 35, or 8 3/4 years.

Programmable calculators have the ability to perform both functions described above automatically. They can execute keystrokes—this is what is going on most of the time in most programs—and they can make decisions. The decision making powers of the calculator reside in its *test keys*: $\boxed{x \geq y}$, $\boxed{x = t}$, $\boxed{\text{INV}}\boxed{x \geq t}$, $\boxed{x = 0}$, and so forth. Notice that you never use these keys outside a program. Their functions are performed in your head when you operate the calculator manually. Within a program, the test keys compare the size of two numbers, and then the program takes either of two courses of action, depending on the outcome of the comparison. The procedure is very similar to your own when you solved the problem above. You mentally compared the contents of the display with the number 2 and then raised or lowered the value of n depending on whether the display was smaller or larger.

Virtually every problem in this book involves the test keys. They give your programmable calculator the appearance of being able to think. It is therefore crucial that you understand exactly how these keys work on your calculator and that you read your manual! Once you understand the test keys, you should be able to solve the following problem.

Suppose that you are in a supermarket confronted with two bottles of ketchup—one containing 24 fl. oz and costing 59¢ and the other containing 28 fl. oz and costing 69¢. Assuming equal quality, which is the better buy? The question can be answered by figuring out the price per ounce (*unit price*) of each item and comparing the two. Hand computations will show that the first bottle costs 2.458 (59/24) cents per ounce and the second 2.464 (69/28) cents. Thus the first bottle is a slightly better buy. Your problem is to automate the comparison on your calculator.

Problem: Write a program that takes as input the price and size (number of units) of two different items of the same product. If the first item is a better buy, the program is to return a 1; if the second item is, a 2. Just to make the problem a little more interesting, have the program return a 0 in the unlikely event that the unit prices are equal.

Solution: We will assume that the price and size of item 1 are stored in m_0 and m_1 and of item 2 in m_2 and m_3. You might store these four numbers by hand before running the program, or you might automate the process of storing data (see Sec. 1.5). Since the numbers to compare are the two unit prices—$u_1 = m_0/m_1$ and $u_2 = m_2/m_3$—the program should begin by computing these two numbers. Once computed, u_1 and u_2 should be loaded into the appropriate registers for testing to see which is larger (or if they are equal). Different calculators use different registers for this purpose. HP calculators use the x and y registers. TI calculators have a special test register, labeled "t". One number is placed in the t-register; the other, in the x-register (the display). We will use the language of the TI. If you have an HP calculator, just substitute "y" for "t" in the following discussion.

This program has three possible outputs: 0, 1, and 2. Therefore, the following three sequences will occur at three different places in the program: a 0 followed by a stop, ($\boxed{\text{R/S}}$, $\boxed{\text{INV}}$ $\boxed{\text{SUBR}}$, $\boxed{\text{GTO}}$ 00, or whatever is most appropriate for your calculator), a 1 followed by a stop, and a 2 followed by a stop. Let's label the sequences a, b, and c for convenience (you may want to label some of them in the program itself if your calculator has labels). Now we want to direct the flow of the program to the appropriate sequence, depending on how u_1 and u_2 compare with one another.

This brings us to the testing functions. Let's postulate that u_1 has been placed in the t-register and u_2 in the x-register. The next step in the program will be the key $\boxed{\text{x} \geq \text{t}}$ ($\boxed{\text{x} \geq \text{y}}$ in the HP). When this step is encountered in a program, the calculator treats it as a question: Is $x \geq t$? If the answer is yes, the program simply continues to the next step. If the answer is no, the calculator skips the next step in the program (or, on some calculators, the next two steps) and jumps to the following step. What should these steps following test $\boxed{x \geqslant t}$ be? Suppose that the answer to the question, x ≥ t?, is no. Since this means that x < t (that is, $u_2 < u_1$), item 2 is a better buy. Hence, sequence c should immediately follow the skipped step(s) so that it will be executed whenever the answer to the test question is "no." Now what if the answer to the test question is "yes"? Then, since you want to send "control" to sequence a or b, the step immediately following the test step should be a GTO order for doing so.[12]

At this stage, you don't know which sequence you want. You know that u_2 is greater than *or* equal to u_1, but you don't know which. Another test is needed to decide. Consequently, the GTO order in the previous paragraph should send the program to a step containing the test key $\boxed{x = t}$ ($\boxed{x = y}$ in the HP). The step immediately following this one (the step reached by a "yes" answer) should be a GTO that sends control to sequence a, and the steps following (those reached by a "no" answer) should be sequence b. The flowchart is shown in Fig. 1-5. [*Solution realized on a TI-58 in 27 steps.*]

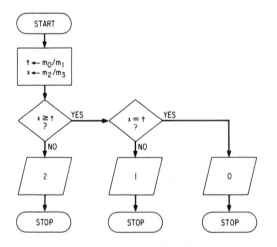

Memory	
0	Price of item 1
1	Size of item 1
2	Price of item 2
3	Size of item 2

Fig. 1-5 Flowchart for test functions

Notes: In this program, the step immediately following both tests was a GTO. Almost all tests work this way; they make the program branch. One branch is reached via the GTO order; the other, which starts immediately after the GTO step, is reached when the GTO is skipped because the answer to test question was "no." On some cal-

---

[12] On some calculators, it is not necessary to use the $\boxed{\text{GTO}}$ key in this step since GTO is assumed by the calculator.

culators (certain TI models, for instance) the step immediately following the test step *must* be a GTO order; the calculator will not accept anything else. On others (the HP-25 and TI-57, for instance) the step immediately following the test can be anything you want. The flexibility can sometimes be used to advantage (see Quickie No. 6 for suggestions).

When we worked this problem on the HP-25, instead of using $\boxed{x = y}$ for the second test, we first subtracted x from y and then used the test $\boxed{x \neq 0}$. The next step was a $\boxed{1}$ and the next 2 $\boxed{\text{GTO}}\,\boxed{00}$. If you think about it a little, you will see that this sequence returns a 0 if $u_1 = u_2$ (x − y = 0) since the $\boxed{1}$ gets skipped and returns a 1 if $u_1 < u_2$ (x − y ≠ 0). As a result, no branching is necessary! Using this trick and the device of loading the original data into the stack instead of memories m_0 through m_3 (see Sec. 1.5), we were able to get the HP-25 program down to 13 steps.

Here is a list of all the different tests that appear in this book (TI owners must substitute "t" for "y"):

x < y	x < 0
x ≤ y	x ≤ 0
x > y	x > 0
x ≥ y	x ≥ 0
x = y	x = 0
x ≠ y	x ≠ 0

Your calculator probably has only some of these tests, but you can generate all the others from the few you have. Suppose, for example, that you have only the tests: x ≤ y and x = y. If you want the test, x ≥ y, simply exchange x and y ($\boxed{x \rightleftarrows y}$ on HP; $\boxed{x \rightleftarrows t}$ on TI); then use the test, x = y. What if you want the test, x > y? Let's say that you wish to execute a sequence A of program steps if this test fails and the sequence B if it passes. You would like the program

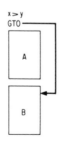

but you have no $\boxed{x > y}$ key. Notice that the test, x > y, fails if and only if the test, x ≤ y, passes. Therefore, the following program does what you want:

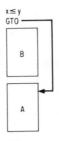

Another approach does not involve switching A and B, as follows:

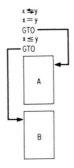

Thus far we have three tricks: (1) interchanging x and y, (2) interchanging blocks of program steps, and (3) breaking one test into two tests. (In the last example, we essentially broke down the single test, x < y, into two: x = y and x ≤ y.) Using various combinations of these three tricks, you can perform any of the six x-versus-y tests listed above. To get the six x-versus-0 tests, just place a 0 in the y register (t register for TI owners).

One final trick: If you have only, say, the test, x < 0, then the test, x < y, can be made by subtracting y from x and testing the result, x − y < 0, using the test, x < 0.

1.4   Writing Programs

Difficulty: 3

The best way to learn how to write programs is to write lots of them, for there is no substitute for practice. One of the main functions of this book, in fact, is to provide you with a collection of problems on which to sharpen your skills. For all that, you will still need expert advice on technique, which often consists of no more than little devices for solving frequently encountered problems as efficiently as possible. The function of this first chapter is to advise you on technique.

General advice is not likely to be of value until you gain some experience in programming and can see how to apply it. It is similar to the value of popular sayings, which frequently seem a little silly to young people but whose meanings deepen with experience. ("Fools rush in" becomes more meaningful the more fools one has known.) But having made this flat declaration, we would like to give you some general advice on writing programs. Don't, however, try to assimilate this material all at once; come back and reread it from time to time as you gain experience in programming. We will begin with a general prescription for writing a program. Then we will let you look over our shoulder as we work out an example. Finally, we will give you a similar problem of your own to work on.

The act of writing a program has three components: (1) finding a method for solving the given problem; (2) turning the method into an actual program for a particular calculator; and (3) editing, correcting, and otherwise cleaning up the program. Although these tend to be three sequential phases, one rarely passes in a straightforward fashion from (1) to (2) to (3). Instead, one passes back and forth between them. You might discover in the course of working out your method, for example, that you need to know exactly how one part of the program is going to look before you can work out the other parts. As a result, you stop and write a little program for that part. Perhaps you then find that things don't work quite as you expected, that you have to go back and modify the method. Or perhaps you see how the method can be simplified as you work on the details. In any case, whatever the programming phase you are in, you should keep the other phases in the back of your mind. Although your attention may be focused on just one of them, in some sense you should be working on all three all the time. Phases (1) and (2) can be described as the big picture and the details, respectively. In solving any mathematical problem, whether or not it has to do with programming, one moves back and forth between these two, never quite letting either one go.

Nevertheless, one should ordinarily start with phase 1, the big picture. Don't rush into the details. Let them slowly begin to form in the back of your mind as you think through a more or less complete idea of the solution. Start phase (1) by fixing firmly in your mind what the program is supposed to find, that is, what is the *output*? Then take a broad overview, letting your mind drift over the various aspects of the problem. What is the input? How are input and output related? Have you seen other problems like this one? If so, can this problem be solved by the same or similar methods? Try to form a mental image of what the program is supposed to do. It may be helpful to draw a diagram. Hold onto your broad overview until you have a pretty complete picture of the solution in your head. Then is the time for phase (2).

Now it is time to get in close. What *quantities* are needed in the various computations? Which ones need to be stored? Keep the *user* in mind. What is the most convenient way to handle input and output from *his* point of view (see Sec. 1.5)? The user should not have to understand the detailed workings of the program in order to use it. Think about *efficiency,* which may lead you back to phase (1). The program probably has to make one or more central computations. If it is a long or complicated one, you should start by writing (and testing) subsidiary programs for these parts. You may want to turn the central computations into subroutines (see Sec. 1.7) and then construct the rest of the program around these central parts. Once your program is completed, write it out in detail. Now you are ready for phase (3), the cleanup.

Begin the process by testing the program thoroughly. Read it through one step at a time and visualize what happens as each step is executed. This can be a tedious job but frequently pays off by uncovering errors, especially if your calculator is new and you are still getting used to it. Now key your program onto the calculator. Choose a selection of inputs for which you know what the output should be, and run these through the program. Sometimes you may not be able to figure out the output in advance, in which case you will just have to check whether it seems reasonable. If the program does different things to different inputs—that is, if it *branches*—be sure to test all types of input. (See Appendix C for a brief discussion of troubleshooting.) Once the program is running properly, you are ready for the final touches. Can it be simplified? Be on the lookout for the same set of steps appearing at two different places; perhaps they need appear only once, as a subroutine (see Sec. 1.7). Are all the steps necessary? It is uncanny how often a step can simply be eliminated without affecting the output. Sometimes a few steps can be saved by rearranging the components of the program.

Finally you have your finished product. Write it out in detail, along with a short description of what the program does and how it works. Include a record of the memories used and their contents. Be sure to include explicit directions for operating the program. This paperwork, called *documentation,* is very important. A good bit of it will seem unnecessary at the time because you have all of it in your head. Two months from now, however, you will not. Nothing is more disheartening than digging out a bare listing for a program you want to use and discovering you no longer understand it. Believe us; this will happen if you don't document your programs.

There is a fourth phase of programming beyond these three: reflecting on what you have done. You can learn by looking back over almost any problem and its solution. Did you acquire any new techniques? Is this problem like others that could be solved by the same or similar methods? Can the program be rewritten to solve a whole class of problems? Try to grasp your solution all at once. You may find that you produced more than you consciously realized while working on the parts; or you may suddenly see a radical simplification of the whole solution, which will lead you, in turn, back to phase 1.

Now we are going to let you watch us solve a programming problem. Imagine that a publisher is currently selling a magazine for $1 a copy. At this price she is able to sell 50,000 copies a month. For every nickel the price is raised, sales will drop by 5000 copies, and for every nickel it is lowered, sales will rise by 5000 copies. What price should she charge for the magazine in order to maximize her income from sales? (This problem is not typical of most in this book because the role the calculator is supposed to play in solving[13] it is not specified. That difference makes it all the more realistic.)

Let's engage in a little preliminary thought (phase 1). From the given information, we can figure out exactly what income can be realized from any assigned price. For example, if the price is raised 10¢ to $1.10, sales will fall to 40,000 copies, and the publisher's monthly income will be 40,000 × 1.10, or $44,000. Since the publisher's current monthly income is $50,000, this is clearly not the solution we are looking for, but it at least makes clear that such a computation can be made for any given price of the magazine. Perhaps we can come up with a formula for the income corresponding to any given raising or lowering of the price, and program the formula onto the calculator. We could then rapidly compare the incomes realizable from various prices. If we want to get fancy, we could even have the calculator do the comparisons using its testing functions. The calculator would then systematically generate all possible raisings and lowerings of the price of the magazine, compute the corresponding incomes from sales, and find and store the price that generates the most income.

Let's see if this scheme can be worked out in detail (phase 2). First, we need a formula for computing the monthly income realizable from a given raising or lowering of the price. You may not have noticed, but a choice has been made here. In general, various quantities can be used as starting points for computing a quantity one is trying to analyze (in this case, income from sales). We have chosen to start

---

[13] In fact, the problem can be solved without a calculator, using calculus (see the Notes).

from the amount by which the price might be raised or lowered, but we could also have started from the price itself. This decision is called choosing the *independent variable*. One wants to select the independent variable in such a way that the formula for the *dependent variable* (income) is made as simple as possible. In our case, the conditions describing the behavior of monthly income are phrased in terms of the amount by which the price might· be raised or lowered. Thus, if we choose this latter quantity as the independent variable, it will be easy to come up with a formula for monthly income.[14]

Let us, then, construct the formula for monthly income from sales. We will assume that the price of the magazine can be raised or lowered only in increments of 5¢. Suppose that the price of the magazine is raised by x increments. The new price of the magazine would then be $(1 + .05x)$. How many people would buy it? Since sales would fall by 5000x copies, the number of copies sold per month would be $50{,}000 - 5000x$. Thus the monthly income realizable by raising the price by x increments—that is, x nickels—will be $\$(1 + .05x)(50{,}000 - 5000x)$. A similar argument will show that if the price is lowered by x increments, then the monthly income will be $\$(1 - .05x)(50{,}000 + 5000x)$. Since the latter expression is the same as $[1 + .05(-x)][50{,}000 - 5000(-x)]$, the same formula will work in both cases if we just put minus signs in front of the x's representing lowerings of the price (that is, if we treat them as negative increases). Thus the formula, $I(x) = (1 + .05x)(50{,}000 - 5000x)$, gives the income obtained from raising the price x nickels, where it is understood that if x is negative, the price is being lowered.

Now it is time for you to start following this discussion on your calculator. Your calculator, of course, can be programmed to turn out values of this formula. (We are assuming that you already know enough about programming to write this type of program. If not, you need the practice.) So go ahead and key the formula onto your calculator (see Quickie No. 7 for a helpful hint). The program should work like this: When x is put in the display and the R/S (or whatever) button is pushed, the calculator returns $(1 + .05x)(50{,}000 - 5000x)$. Test your program by trying it out for $x = 0$ and $x = 2$ (the latter corresponding to a price increase of 10¢). We already know that these values should return 50,000 and 44,000, respectively. By experimenting with various values of x, you will discover that the biggest return is generated when $x = -5$. Thus the maximum possible income will be generated by lowering the price five increments to \$.75. The corresponding monthly income is \$56,250.

---

[14] There are times, however, when a seemingly unnatural choice of the independent variable can allow one to see through a formula that looks very complicated in its original form. This is the art of *changing variables,* which plays a large role in mathematics, especially in the theory of equations.

For this particular problem, what we have done so far probably represents the best use of the calculator. Once we had a formula for computing the monthly income corresponding to any given raising or lowering of the price of the magazine, we programmed the formula onto the calculator. It was then a relatively easy matter to find the best price for maximizing monthly income by simply trying out various values of x. In other problems of the same type, however, there may be too many possible values of x for this to be a practical approach. We are now going to show you, with the same problem, how the calculator can be programmed to conduct a search for the optimal x.

Let's be as clear as possible about what we want to do. Whenever the calculator takes over a new part of a problem, the details of that part must be made explicit. Here is the situation. We have a formula, $I(x) = (1 + .05x) (50,000 - 5000x)$, in which x represents the number of increments by which the price of the magazine might be raised or lowered, and $I(x)$ represents the monthly income realizable by raising or lowering the price by x increments. We seek the value of x that will make $I(x)$ as large as possible.[15]

First, let's see within what range of values the x we are looking for must fall. Notice that if the price of the magazine is raised 10 increments (that is, by 50¢), sales will fall by 50,000 copies (that is, to 0). Obviously, the optimal value of x must be less than 10. On the other hand, if the price is lowered by 20 increments, the magazine will be selling for nothing and will generate no income. Thus the optimal x must also be larger than -20. What we would like to do is have the calculator systematically generate all values of x between -20 and 10, compute the corresponding values of $I(x)$, and find the x that generates the largest value of $I(x)$. We cannot generate all possible values of $I(x)$, store them, and *then* compare them because there are too many. Rather, we will somehow have to make comparisons as we go along. Can you think of an analogous situation in which you try to pick the best from among a large collection of objects that cannot be examined all at once?

Suppose that you are at a sale table piled with pairs of socks, and you want to buy just one pair. How do you find the pair you like best? One way is the following: You pick up one pair more or less at random and hold it in your hand. Then you start going through the pile. As you look at each pair of socks, you compare it with the pair in your hand. Whenever you come across a pair you like better, you exchange it with the pair in your hand. Continuing this way until you have gone through all the socks, you buy the pair you hold in your hand at the end.

---

[15] Of course, we already know what the answer is: $x = -5$. The following analysis, however, is based on the assumption that the problem is not yet solved.

We are going to have the calculator do a similar search. First, an arbitrary x and I(x), say 0 and 50,000, will be stored in a pair of memories (which will play the role of your hand in the example above). Next, the calculator will generate all possible x's and I(x)'s one at a time, starting with x = −19. As each new x and I(x) are generated, the new I(x) will be compared with the one in memory. If the new I(x) is larger, it will replace the one in memory, and the new x will replace the x in memory. At the end of the routine, when all the I(x)'s have been examined, the I(x) in memory will be the biggest possible, and the x in memory will be the x that produces it.

Four memories will be used. To begin, m_0 will store the impossible low value of x (in this case, −20) and will successively store all possible values of x: −19, −18, −17, . . ., 8, 9, and 10. Also, m_1 and m_2 will store the best values of x and I(x) found so far as the program proceeds, beginning with 0 in m_1 and 50,000 in m_2; at the end of the routine, the values of x and I(x) we are looking for will be in these two memories. Finally, m_3 will be used to store the impossibly large value of x (in this case, 10; the technical names for − 20 and 10 in this setting are the *lower* and *upper bounds* on the solution). After each value of x has been generated and stored in m_0, it will be compared with the contents of m_3. As soon as the contents of m_0 equal or exceed the contents of m_3, the routine will be terminated.

The program itself will work as follows: Increase the contents of m_0 by 1 and check to see if the result is greater than or equal to the contents of m_3. If it is, stop; the routine is over. If not, take the new contents of m_0 and compute I(x) from it [the program you have already written for computing I(x) will be embedded within this larger program]. Next, check to see if the resulting I(x) is larger than the contents of m_2. If it is, store the new I(x) in m_2, and store the contents of m_0 in m_1. These are the new best values of x and I(x). If the new I(x) is not larger than the contents of m_2, skip the two storages; the values in m_1 and m_2 are still the best yet. Finally, loop back to the second sentence of this paragraph. The flowchart for this program is shown in Fig. 1-6.

Problem: Suppose that a manufacturer wants to build a tin can with a capacity of 1 liter. What should the dimensions of the can be in order to minimize the amount of sheet metal used in its construction? This question is equivalent to asking for the dimensions of a cylinder with a volume of 1 liter and minimum possible surface area. Construct a calculator program for finding the dimensions of that optimal cylinder. Arrange it so that your answer is correct to the nearest millimeter (tenth of a centimeter). Here are some hints: The volume formula for a cylinder is: $V = \pi r^2 h$, where r is the radius of the base of the cylinder

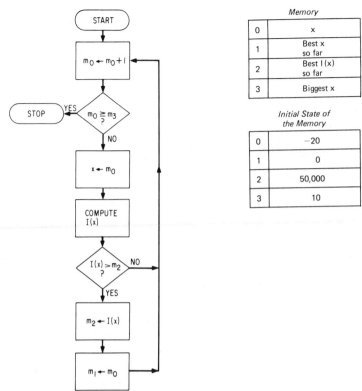

Fig. 1-6 Flowchart for comparison search [maximizing I(x)]

and h is the height. If r and h are measured in centimeters, then V will be given in cubic centimeters. One liter is 1000 cubic centimeters. The formula for the surface area of a cylinder[16] is: $S = 2\pi r^2 + 2\pi rh$.

Solution: Because of the strong similarity between this problem and the magazine problem, one might hope that the same basic scheme would work. We are looking for those dimensions of a cylinder of given volume that will minimize its surface area. The idea will be to generate various dimensions systematically, compute the corresponding surface areas, and pick out the best of them—using the method described in the example. The first thing we need is a formula for computing the surface area of a cylinder from its dimensions.

Here we run into a problem. The formula, $S = 2\pi r^2 + 2\pi rh$, has two independent variables: r and h. To generate pairs of variables

---

[16] The expression $2\pi r^2$ is the area of the base and top of the cylinder, and $2\pi rh$ is the area of the side. If you imagine taking the top and bottom off the can, cutting it in a vertical line down the side, and then flattening the side out, you will get a rectangle whose base is the circumference of the base of the cylinder ($2\pi r$) and whose height is h. The area of this rectangle is thus $2\pi rh$.

systematically is a tricky business involving nested loops (see Sec. 1.8). Moreover, we are not interested in just any old r and h. We want only those for which the value of $\pi r^2 h$ (the volume of the can) is 1000. Actually, this restriction provides the way out of the problem. Since the volume of the cylinder we are looking for is specified, once the radius of the base has been given, the height will be forced to be exactly that number h such that $\pi r^2 h = 1000$. This equation can easily be solved for h: $h = 1000/\pi r^2$. This value for h can in turn be plugged into the formula for S, allowing us to write the latter in terms of the single variable r: $S = 2\pi r^2 + 2\pi r(1000/\pi^2) = 2\pi r^2 + (2000/r)$. This is the formula we need.

We can now generate a sequence of values of r, computing the corresponding values of S as we go, and find and store the smallest value of S and the r that produces it. Once r has been determined, we can find the height of the cylinder from the formula: $h = 1000/\pi r^2$. Almost exactly the same routine used to solve the magazine problem can be used here. The subprogram for computing $I(x)$ will be replaced by a subprogram for computing $S(x)$: $S(x) = 2\pi x^2 + (2000/x)$. Once again we leave it to you to construct this program.

Two other modifications of the previous program are required. In the magazine problem, we were trying to *maximize* $I(x)$. This time, since we want to *minimize* $S(x)$, we need to find and store that x which yields the *smallest* value of $S(x)$. Look back at the flow chart in Fig. 1-6. To find the smallest value of $S(x)$, all we have to do is change the previous test, $I(x) > m_2$, to obtain the required test, $S(x) < m_2$. Second, in the magazine problem, we used nickels as units and were interested only in whole numbers of these units. Thus, the contents of m_0 were incremented by 1 each time through the loop (see the top box of the flowchart in Fig. 1-6). This time we use centimeters as units, but since we want to find the answer to the nearest tenth of a centimeter, we will want to increment m_0 by .1 on each pass through the loop. We accomplish this by changing the top box of the flow chart from $m_0 \leftarrow m_0 + 1$ to $m_0 \leftarrow m_0 + .1$. These two changes yield the flowchart shown in Fig. 1-7.

Finally, we need to find the numbers to put in memories m_0–m_3 before the program is started. The number in m_0 should be a *lower bound* on the solution, a number that we know to be smaller than the number r being sought. A convenient number to use in this case is 0. Clearly, the radius of the optimal cylinder (of *any* cylinder, for that matter) is a positive number (that is, larger than 0). An *upper bound* on the solution has to be stored in m_3. This is a little trickier to find. One might call 0 a *natural* lower bound in this problem since there are simply no cylinders with dimensions less than 0, but since we can imagine cylinders with dimensions as large as we like, we are going to have to find the upper bound experimentally. Luckily, we need

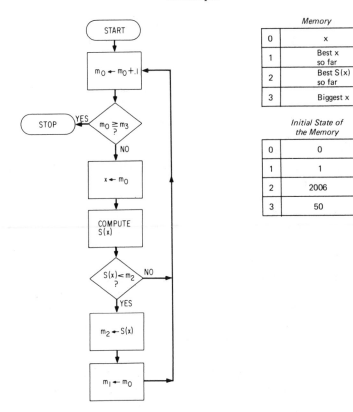

Fig. 1-7 Flowchart for comparison search [minimizing S(x)]

perform the experiment only in our heads (and with the help of our calculators). Imagine a cylinder with a really large base, say a tin can with a diameter of 1 meter. If this cylinder is to have a volume of 1000 cu cm, then, since its radius r is 50 cm, its height must be

$$h = 1000/\pi r^2 = 1000/2500\pi = 0.127 \text{ cm}$$

It seems intuitively clear that a tin can that is 1 m wide and about 1/8 cm high is not the most efficient possible. The top and bottom of the can will require more than a square meter of sheet metal. If we let the radius of the base get larger, moreover, the situation will just get worse. The optimal radius, then, must be less than 50 cm. We could obviously improve on this upper bound, but why bother? We are supposed to be getting the calculator to do the work. We therefore put 50 into m_3.

Now how about m_1 and m_2? These are the memories in which the optimal x and S(x) will be found after the program has been run.

We need to put something in them for comparison purposes when the program starts. Let us select an arbitrary x, say 1; then,

$$S(1) = 2\pi(1^2) + (2000/1) = 2\pi + 2000 \doteq 2006$$

Store these numbers in m_1 and m_2, respectively. The program is now ready to run. It should finish with the optimal values of r and S in m_1 and m_2, respectively: r = 5.4 cm and S = 554 sq. cm. [*Solution was realized on a TI-57 in 32 steps.*]

The above program has one drawback. It is rather slow, taking about 15 minutes to run on a TI-57. What makes it slow is the number of x's that the program has to check out. They range all the way from 0.1 to 49.9 in increments of 0.1—a total of 499 numbers to check. Things can be speeded up considerably if, instead of trying to get the answer to the nearest millimeter on the first run, we zero in on this level of accuracy in two stages. If the step at the top of the flowchart is changed back to its original form, $m_0 \leftarrow m_0 + 1$, the program will then find the optimal radius to the nearest centimeter, or 5, after checking only 49 numbers. Since the optimal whole number radius is 5, the radius we are looking for should be somewhere between 4 and 6.[17] We therefore put these numbers in m_0 and m_3 and change the first part of the program back to its amended form, $m_0 \leftarrow m_0 + 0.1$ (leaving the contents of m_1 and m_2 as they are.) Now when we run the program, the desired answer is achieved after a check of only 19 numbers. This approach is much faster, for the calculator has to check a total of only 78 numbers.

It is a little inconvenient to have to change the program for the two runs, but an easy remedy is available. All we have to do is set up a memory, say m_4, to store the size of increments that we want to use in given runs of the program and then store those increments before running the program. The first part of the program now takes the form, $m_0 \leftarrow m_0 + m_4$. The flowchart for this improved program is shown in Fig. 1-8.

Now it is time to take a closer look at what we have done. It should be pretty clear that we have evolved a general technique for finding the maximum or minimum value of a mathematical expression with one independent variable. Suppose that Q is some quantity given by a formula in terms of x, for example, $Q = 2\pi x^2 + (2000/x)$. We wish to find the value of x which minimizes Q. First we must find a pair of *bounds on the solution,* a number x_ℓ (the *lower bound*) that we know is less than the number we are looking for and a number x_u (the *upper bound*) that we know is larger. Then we take an arbitrary x between x_ℓ and x_u and compute Q(x), the corresponding value of Q. These four

---

[17] This reasoning doesn't always work (see the Notes).

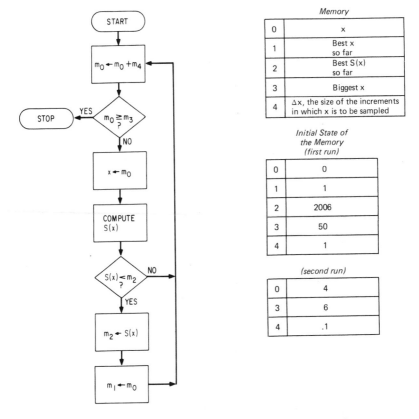

Fig. 1-8 Flowchart for minimizing S(x)—speeded up version

numbers are placed in memory. We also store Δx, the size of the increments in which we wish to sample the numbers between x_ℓ and x_u.

The program generates the numbers between x_ℓ and x_u one at a time in increments of size Δx. As each number x is generated, Q(x) is computed. A program for computing values of Q from x will be embedded in the larger program. If your calculator has subroutines, the program for Q should be a subroutine, because it will then be easier to change for different Q's. As each value of Q(x) is computed, it is compared with the value in memory. If smaller, this value of Q(x) and the x which produced it replace the values in memory. As each new x is generated, a check is run to see if $x \geq x_u$. As soon as this x occurs, the program is terminated.

The flowchart for this routine (Fig. 1-9) looks exactly like the modified flowchart for minimizing the surface area of a 1-liter cylinder. To obtain a flowchart for *maximizing* Q, merely change the test, $Q(x) < m_2$, to the test, $Q(x) > m_2$. The skeleton program (without a

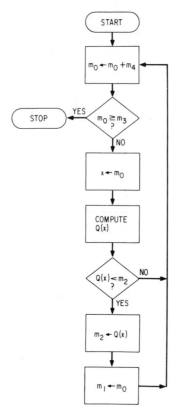

Memory	
0	x
1	Best x so far
2	Best Q(x) so far
3	x_u
4	Δx

Initial State of the Memory

0	x_l
1	Arbitrary x
2	Value of Q at the x in m_1
3	x_u
4	It's up to you.

Fig. 1-9 Flowchart for minimizing the quantity Q given by a formula with one independent variable

program for computing values of Q) can be programmed onto a TI-57 in 18 steps.

Here are two other problems for you to solve with this routine:

1. You have a piece of cardboard 12 cm square and want to make an open-topped box out of it by cutting squares from the corners and folding up the resulting flaps. (see Fig. 1-10(a)]. What size squares should you cut from the corners to maximize the volume of the box?

2. A farmer has 100 feet of fencing. He is going to make a pig pen with it by enclosing a rectangular plot next to his barn. The side of the barn will form one side of the rectangle, and the fence will form the other three sides [see Fig. 1-10(b)]. What dimensions should the rectangular plot have to maximize its area?

Notes: These two problems, like the two preceding them, are ex-

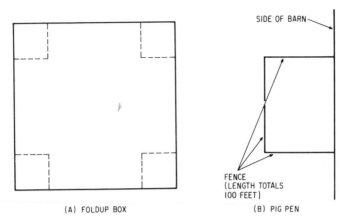

SIDE OF BARN

FENCE
(LENGTH TOTALS
100 FEET)

(A) FOLDUP BOX   (B) PIG PEN

Fig. 1-10 Diagrams for fold-up box and pig pen

amples of problems in *optimization*, a large and important topic in mathematics. Optimization problems, which appear in both pure and applied mathematics, have played an important role in the history of mathematics. Problems of the type given here spurred Fermat to create the fundamentals of differential calculus before the work of Newton and Leibniz.

Optimization problems can be very difficult. Although computers have greatly extended our ability to solve them, there are still some whose solutions lie beyond the capabilities of even the biggest computers. An example is the so-called traveling salesman problem. A traveling salesman has a territory that covers, say, 25 towns. He needs to visit each town once each month and knows the distance by road between any two of them. His problem is to find the circuit of minimum length that passes through all the towns. The problem obviously has a solution, since there are only a finite number of possible routes. Nevertheless, that number is incredibly large. Even in this relatively simple case, there are 6.2×10^{23} possibilities! If a computer were able to check out one billion of these possibilities per second, it would take more than a million years to try them all.[18] Many practical problems—for example, airline routing problems—turn out to be of this type.[19]

In the problems we have worked on here, the number of possibilities to be checked were relatively few. Moreover, we developed a

---

[18] No existing computer can run a billion checks per second.

[19] The traveling salesman problem belongs to a class of problems called *NP-Complete*. It has not been proved that there isn't a more efficient algorithm for solving such problems, but it is strongly suspected that no shortcuts are possible.

strategy for avoiding having to check them all (the strategy of zeroing in on the level of accuracy desired). Calculus also provides a strategy for solving these four problems. It turns out that the optimal values of the quantities involved in all of them occur at the points at which the derivatives of the functions describing the quantities are zero. If you have taken calculus, you have undoubtedly used this strategy to solve optimization problems of the same sort. In problems of the traveling salesman type, however, most computer scientists believe that there simply is no strategy for quickly zeroing in on the optimal solution. Since we are pretty much stuck with trying all the possibilities, the problems are for all practical purposes unsolvable. Even if we happened to find the optimal solution by chance, we would have no way of knowing that there was not a better one.

Notice that the diameter and height of the optimal 1-liter can turned out to be about equal (if they had been computed more accurately, they would have turned out to be exactly equal). In the magazine problem, moreover, the optimal price of 75¢ resulted in a sales of 75,000 copies. Although conditions of "equality" of this type frequently characterize optimal solutions, they are not always so simple. The analogous condition in the cardboard box problem sounds artificial if you don't know calculus. See if you can find the "equality" condition in the optimal pig pen problem (the optimal plot is 50 by 25 ft).

We should mention two limitations of the optimization routine described above. One has to do with the strategy of zeroing in on the desired level of accuracy. If the graph of the function to be optimized contains quick jumps in value, the best whole number solution may not be close to the best solution that has one digit after the decimal point. Thus you might find yourself zeroing in on the wrong place. This situation is not likely to occur unless your formula for Q is fairly exotic, but it is something you should keep in the back of your mind. Another limitation concerns the level of accuracy with which the optimal x can be determined. As you get close to the optimal value for the quantity Q, the graph of Q tends to flatten out (a characteristic related to the fact that the derivative, or rate of change, is 0 at the optimal values). Once you get down to a certain level of accuracy for x, the corresponding values of Q will be indistinguishable on the calculator. Try calculating cos(0) and cos(0.001) on your calculator, for example. We know that cos(0) is bigger, but your calculator cannot tell the difference. Thus, beyond a certain level of accuracy, the calculator will not be able to pick the truly optimal value of x. In a way it doesn't matter, since what you are really trying to optimize is Q, and the x that the calculator finds will yield a Q that is virtually indistinguishable from the correct value.

1.5   Loading Data and Other Aesthetic Considerations

Difficulty: 2

Almost all calculator programs require the user to supply numbers for the program to operate on. These numbers are referred to as the *input* or *data* for the program. The program *processes* the data and returns *output*. The input and output constitute what is called the *man-machine interface*. They are the language in which the calculator and its user talk to one another. In this section, we will discuss various ways of supplying the calculator with data and the merits of each.

There are basically two places into which data can be entered: the display or a memory. Entering data in the display is easier on the user and less likely to generate errors (he might put the data in the wrong memory), whereas entering data in a memory is generally easier on the calculator. More about this later. First let's look at techniques for loading data through the display. The idea here is to automate the program as much as possible: The user of a program should not have to understand how it works in order to use it. All he should have to do is enter the data, run the program, and get the desired output.

Let's consider a straightforward example. You are asked to produce a calculator program for finding the height of buildings with a barometer. Your solution is the following: the user is to drop the barometer off the top of the building to be measured and measure the length of time it takes to hit the ground. If t is the length of time it falls, then the distance it has fallen, s, is given by the formula,

$$s = 16t^2$$

where s is measured in feet and t in seconds.[20] A program is required that takes t as input and returns s as output. You want to set the program up so that the user enters t in the display and hits the run button, and the calculator returns the value of s. The flowchart for this program is shown in Fig. 1-11.

Now let's consider a little more complicated example. A college admissions board uses the following scheme for rating its applicants. After his application form has been analyzed, each applicant is assigned ratings in the following four areas: (1) scholastic aptitude, (2) scholastic achievement, (3) psychological stability, and (4) the amount of money his parents donated to the school the previous year.

---

[20] This formula ignores the effect of air resistance. For a discussion of this factor, see a text on differential equations.

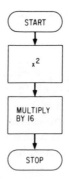

Fig. 1-11   Flowchart for building height program

Each of these ratings is a number between 0 and 100. The board has decided that the relative importance of the four factors are 15, 15, 30, and 40 percent, respectively. An applicant's "net" rating is obtained by multiplying rating (1) by 0.15, rating (2) by 0.15, rating (3) by 0.3, rating (4) by 0.4 and then adding up the resulting numbers. A calculator program for making this computation is desired.

The program will require four pieces of data, but how should this input be handled? The program operator could put the four numbers into four separate memories, but it would be easier to have the program process the data one piece at a time, stopping at intervals to enter the next number. The flowchart for the basic program will then look like the one in Fig. 1-12.

This program can be improved. If you have ever used a similar one, you know that it is easy to forget which numbers have been entered. To avoid this problem, the program can be constructed to give *cues* indicating which numbers have already been processed. For example, after storing the product of 0.15 times the first rating in m_0 and before the first R/S, a 1 can be put in the display to indicate that the first number has just been processed. Similarly, a 2 can be entered in the display just before the second R/S, and so forth.[21]

Every approach to a problem has associated costs that must be weighed against associated benefits. The price usually paid for programs that make things easy on the user is the increase in their length. The preceding program takes 34 steps on a TI-57, a calculator with only 50 steps available. If the problem were more complicated, the program could easily overrun the capacity of the program memory. Suppose, for example, that the admissions board hired a consulting firm to revise its admission criteria and the firm recommended use of the following expression for the net rating of each applicant:

---

[21] Be careful when using this trick that the cue does not become an unwanted digit in some other number (see the Notes).

$$\frac{\sqrt{r_1{}^2 + r_2{}^2}}{\arctan\left(\dfrac{4}{3}\,\pi r_2{}^3 + r_1{}^2 - 2r_3r_4 + r_2{}^2\right)}$$

The overriding concern would now be how to get a program for this formula onto the TI-57.

For such programs, it is usually more efficient to load the data into memories. Besides eliminating the cueing and stopping steps, this arrangement allows the data to be processed more conveniently since the calculator no longer has to wait for the user to present it piece by piece. In the previous program, for example, since the first two ratings are both to be multiplied by 0.15, it would be more efficient to add them first and then multiply the sum by 0.15. In programs in which a given piece of data is to be used more than once, space is always

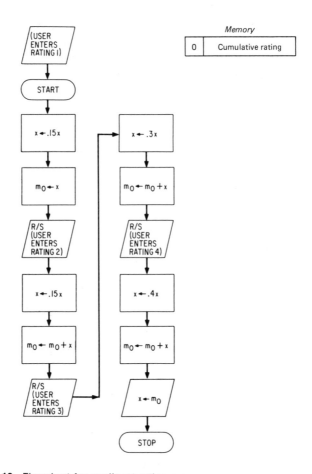

Fig. 1-12 Flowchart for applicant rating program

saved if it is loaded into a memory, for if the operator does not put it in a memory, the program must.

Saving space in a program is of greater concern with small calculators than large ones, but it is good mental discipline whatever the size of your calculator. The best programs are those exhibiting both virtues—user convenience and economy of steps—to the degree appropriate for the problem to be solved and the calculator used to solve it.

There is, however, an aesthetic consideration that runs counter to both these values, namely, the intelligibility of the program. A program with a lot of little extras tacked on for the convenience of the user and a lot of little space-saving tricks worked into it as well is likely to be very difficult for another person to understand without outside explanation. When you write such a program, you would be well advised to write an accompanying explanation of what it does and how it does it. Two months from now, *you* will be that other person trying to make sense out of your own program. On the other hand, there is nothing more pleasant to read than a program that solves a problem elegantly and at the same time explains itself as it proceeds. When you can produce programs that are convenient to run, economical of space, and instructive to read, you may consider yourself a master in the art of programming.

Here is a problem for you to sharpen your skills on.

Problem: If a projectile is fired from ground level at an angle of α to the horizontal and with an initial velocity of v_0 ft/sec its height off the ground after t seconds will be given by the formula,[22]

$$h = v_0 t \sin \alpha - 16t^2$$

and its distance downrange (horizontal distance from its starting point) by the formula,

$$d = v_0 t \cos \alpha$$

Write a program that takes v_0 and α as input and returns as output the time and distance downrange at which the projectile strikes the ground.

Solution: The projectile will strike the ground when h = 0. Factoring the equation,

$$v_0 t \sin \alpha - 16t^2 = 0$$

we get

$$t [v_0 \sin \alpha - 16t] = 0$$

---

[22] Once again we are ignoring air resistance.

which yields t = 0, the time when the projectile leaves the gun, and t = v_0 sin α/16, the time we are looking for. Plugging this value for t into the formula for d, we see that the projectile will be $v_0{}^2$ sin α cos α / 16 feet downrange when it hits the ground. The program, then, needs to compute the two quantities, t and d.

To make the program easy on the user, we will enter the data through the display. On TI (algebraic) calculators, v_0 will be entered first. The program will begin processing v_0 and will stop as soon as α is needed to continue. Since v_0 and α appear more than once in the formulas, they apparently also need to be stored as soon as they have been entered. The TI program might begin by computing v_0 (sin α)/16. If so, the first few steps of a TI-57 program would look as follows:

$\leftarrow$ (user enters V_0)
STO 1
×
R/S
$\leftarrow$ (user enters α)
STO 0
sin
÷
|
6
=

The program has now computed the desired value of t and contains α and v_0 in storage ready for further computations.

The HP (RPN) program doesn't need to stop in midstream like the TI. Instead, the user can enter both v_0 and α into the stack at the beginning. The program can then go on its own way to compute t and d. To make the output easy to retrieve, we will conclude the program with it in the x and y registers (TI: x and t registers) so that the user can flip back and forth between the two outputs simply by pressing x ⇌ y (TI: x ⇌ t).

As a final convenience, the program should end in such a way that it is all set to be ru   a second time. On the HP-25 this is easy; we just finish with GTO 00. This sets the program pointer back to the top and stops the program. On a TI-57, the program should end with R/S followed by RST. This sends you back to the top of the program as soon as you press R/S for the next run. If your calculator has subroutines, you could label the whole program a subroutine and simply call for it as such whenever you need it.

So much for user convenience. Now how about space? Very little space can be saved by sacrificing convenience in this program. Two STO orders can be eliminated if the user stores the data by hand, and a step or two might be saved by not worrying about running the program a second time, but the loss of convenience makes these savings not worth it. The real space-saving tricks in this program come from being mathematically observant. Notice that the distance d we are looking for is v_0 (cos α) multiplied by t, where t = v_0 (sin α)/16. Consequently, t should be computed first, and d obtained from it by multiplying by v_0 (cos α). The only thing you have to watch here is that you don't lose t when you use it to compute d. The TI-57 program we started previously can be continued as follows:

$$
\begin{aligned}
&\times \\
&x \rightleftharpoons t \\
&\text{RCL I} \\
&\times \\
&\text{RCL 0} \\
&\cos \\
&=
\end{aligned}
$$

You now have t in the t register and d in the display. Notice how t was flipped out of the way with the $\boxed{x \rightleftharpoons t}$ key just before being eaten up by the multiplication. (See Quickie No. 10 for a further discussion of this type of trick. On HP calculators, you can use the "last x" key to recover t after the multiplication.) This trick produces a fairly concise (and fairly readable) program that still offers all the user conveniences mentioned in the previous paragraph. Our TI-57 program took 19 steps, and our HP-25 program took 15. If you produce a program with these specifications, give yourself a B+.

Another level of sophistication can be reached by the programmer who knows the keys and their functions. If you put v_0 and α in the x and t registers (HP: x and y registers) and press the key for conversion from rectangular to polar coordinates, the calculator will return to you, free of charge, v_0 (cos α) in the x register and v_0 (sin α) in the t register.[23] With these numbers in hand, you are a long way toward producing t and d. [*Solution was realized on an HP-25 in seven steps; on a TI-57 in nine steps.*]

Notes: When using a cue, you must be careful that it does not become an unwanted digit of the number keyed in when the program stops. If 1 is used for a cue and the program goes . . . , 1, R/S;—for example— the display may read 15 when the user keys in a 5. (Check it out on your own calculator.) The problem can occur in other situations, too, and can be murder to uncover when your program doesn't work. It can be avoided on HP calculators by inserting an ENTER after the 1, and on TI calculators by inserting an equals sign.

In addition to giving cues *to* the user, programs can be written

---

[23] See Sec. 2.12.

to accept cues *from* the user. Say your program contains two sequences, s_1 and s_2. Sometimes you want the calculator to execute sequence s_1 on a number entered in the display, and sometimes, sequence s_2. Cues *from* the user permit the calculator to decide which sequence you want. One quite useful and economical cue is to put a minus sign on the numbers that are to have sequence s_2 executed on them. A test, $x < 0?$, can then be used to send control to the appropriate sequence. Particular numbers like 1, 2, and 3 can also be used as cues.

If program space is of prime importance, you may want to "preprocess" some of the data before putting it in a memory. In other words, you perform some of the work that the program would otherwise have to. In less critical situations, the program may be able to save space by preprocessing its own data. Suppose, for example, that a program takes x and y as input and that these variables only appear in the expression $\sqrt{x^2 + y^2}$. It will then be more economical to compute $\sqrt{x^2 + y^2}$ at the beginning of the program and store *it* for future use, forgetting about storing x and y separately.

Although the subject has been mentioned, we have not said enough about using the stack for loading data if you have an RPN calculator. The stack is intermediate between the display and the memories, sharing some of the virtues of each. It is easy to load data into one. You need only use the ENTER key. On the other hand, the stack functions like the memories in that data placed there can be juggled conveniently. (See Quickie No. 10 for further remarks about the stack.)

Finally, we have said nothing here about two major devices for loading data that are available to owners of large calculators: user definable keys and data cards. User definable keys can be very convenient for this purpose. A subroutine under a given label can be set up whose only function is to store the number in the display in a particular storage register. The name of this number can then be written on a little card that is slipped into the face of the calculator as a cue to the user. User-defined keys can also be employed to preprocess data, making the main program less complicated.

If you own a calculator with magnetic card storage, these cards can be used to hold large data sets that will be put back into the calculator later for more processing (say, your checkbook records). Data cards can also be used to expand programs beyond the apparent capacity of your machine. You may be able to break a long program into parts that can be stored on more than one magnetic card. The first part does its thing and captures the result in the memories. The contents of the memories are then dumped onto a data card while the second part is loaded into program memory. The data is then loaded back, perhaps after other processing is done, for further processing.

1.6   Output

In the last problem we were concerned with input—how the user tells the calculator what he wants it to know. Now we turn to the other end of the process, output—how the calculator imparts its answers to the user. Since input and output are the phases of computation in which the user and the calculator are in direct contact, one should be guided by the user's viewpoint in working on them.

Working on output is often a matter of developing little touches to make things more convenient for the user. It is preferable, for instance, to have the output appear in the display rather than in a memory, from which it will have to be eventually fished out. If the output consists of a pair of numbers, as in the previous problem, a nice device is to have it end in the x and y registers (x and t registers on some calculators) so that the user can see one number and then the other by pressing $\boxed{x \rightleftharpoons y}$ (or $\boxed{x \rightleftharpoons t}$). You may also want to attatch cues to the output. If a program is going to output a series of numbers, it may be helpful to flash an identifying number—1, 2, 3, . . . —before each so that the user will know where he is in the series. If some output is particularly significant, you may wish to emphasize it by flashing it for more than one PAUSE or by flashing a minus sign on and off in front of it.

From a larger point of view, what we are trying to do is paint a picture with the output. Every program is devised to communicate something, the input and output being the medium. Communication is an art, not a science. There is room for real skill and creativity here. For instance, have you ever thought of doing something with the two digits on the right in the scientific notation format, or with the parts of a number that lie to the left and right of the decimal point? (See Sec. 1.14).

The output format should enhance the message being conveyed through the numbers. How much detail do you want to provide? More detail usually means more numbers or more places after the decimal point. At times, accuracy demands great detail. At other times, it leads only to confusion. Who can assimilate all ten digits of a number that is flashed in the display for a fraction of a second?

An important parameter in any program is speed, and output format is frequently the controlling factor in how fast or slow a program will run. Suppose your program is going to produce a series of numbers leading toward some end result. You should ask yourself if you want the user to see these numbers. Will they promote understanding? Is it just a matter of watching the program at work for its own sake? Or does the user *need* to see the numbers to know whether or not to stop execution? If you are going to show the numbers to the user, you need to decide whether to do so with an R/S or a PAUSE.

The PAUSE is faster, but R/S allows the user to digest the numbers at his own pace. Perhaps the program can be made flexible, allowing the user to see the numbers at some times, and at other times, when speed is most important, dropping the PAUSEs altogether (see Ulam's Problem in Sec. 2.3 for an example).

Now let's look at a concrete situation. Whenever a computational process is applied repeatedly to its own output, we say that the process is being *iterated*. Compound interest, for example, is an iterative procedure. Let's say that you have $100 in a savings account that pays at an annual rate of 8 percent, compounded quarterly. The $100 can be viewed as the initial input. At the end of each quarter, the money in your account is multiplied by a factor of 1.02; thus at the end of the first quarter you will have $102 in your account. This amount is the *output* for the first quarter. The computational process is multiplication by 1.02. In the next quarter the $102 becomes the input, and the process is applied again, producing $104.04 in your account. This output from the second iteration becomes the input for the third iteration, which produces the third quarter's balance of $106.12, and so on.

Iterative procedures are used to model a wide variety of phenomena. For example, economists view a nation's economy as an iterative procedure. The economy is described using a number of variables: gross national product, rate of inflation, rate of unemployment, etc. These variables are thought to be both a description of the present state of the economy and the determining factors for the future state of the economy (in other words, the output of the present state of the economy and the input which will determine the next state of the economy under the action of whatever economic laws operate in the nation). Strict determinists would claim that the universe is an iterative procedure. Each state of the universe is the product of its previous state under the operation of the unchanging laws of nature, and these same laws will produce the next state from the present one.[24]

A question frequently asked of iterative procedures is: Is there a stable state? That is, is there an input that produces itself as output? And if so, what inputs will eventually lead to the stable state? In the compound interest example, there are no stable states; each input produces an output larger than itself. Economists would be very interested to find a stable-state economy; there are no examples in the modern industrial world. A stable state of the universe would be very *un*interesting (*nothing* would change). However, astronomers still debate whether the universe is *cyclically* stable (that is, will the present state of the universe recur at some, perhaps remote, future time?).

---

[24] This view of the world has been largely abandoned by physicists in the twentieth century.

Now let's bring this discussion down to the dimensions of a calculator.

Problem: You have a computational procedure that can be programmed on a calculator, for example, $f(x) = \sqrt{x + 1}$. The problem is to write a program around the program for the computation, the function of which is to iterate the computation and search for stable states. We will assume that there is no theory for finding inputs that lead to stable states. The user is simply to choose an initial input and start the program, which will then discover whether or not the input leads to a stable state. Remember that this problem is supposed to be concerned primarily with output. Think about how to make the output do the most for the user.

Solution: There is one very simple solution. The program for the computational process is followed by an R/S to allow the user to see the output. This step is followed by a GTO order sending control back to the beginning of the computation. When R/S is pressed a second time, the program will loop back and run the computation again on the contents of the display, which will be the output from the previous run (see Fig. 1-13).

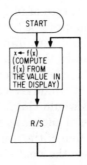

Fig. 1-13　Flowchart for stable state program

The user places his first input in the display and presses R/S. The calculator stops with $f(x)$ in the display. He presses R/S again and gets the next iteration. He then keeps pressing R/S and watches the display to see if the values are settling down toward stability. If your calculator has subroutines, the same thing can be accomplished by programming the computation as a subroutine (with a user-defined key if you have one). Each run of the subroutine will then produce an iteration of the computation, provided that you leave the results of the last run in the display.

Try this routine on the equation, f(x) = cos x, with the calculator in degree mode. The program will be as follows:

01	cos x
02	R/S
03	GTO 01

You will see that no matter what value you start from, the
value in the display will quickly settle down to .999847742. Now try
the same routine on the equation, $f(x) = \sqrt{x + 1}$. Here the stable value
is 1.618033989, but the program doesn't find it so quickly. You can
save the user the trouble of repeatedly pressing R/S by replacing it
with a PAUSE (see Fig. 1-14). Now all the user has to do is watch the
display and stop the program when a stable value has been reached.

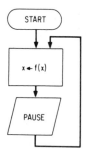

Fig. 1-14 Flowchart for stable state program (first revision)

Even this approach can get boring, however. Let's see if the
program can be speeded up even more by eliminating the PAUSE. To
do so will require the calculator to take over the former function of
the user, that of watching for the stable value to appear. The idea will
be to make the calculator store the values produced by the computation
and compare these with their successors, stopping as soon as equality
occurs. The flowchart for this approach is shown in Fig. 1-15).

This program will run itself until a repeating output is found,
but it contains a rather subtle flaw in the test, $x = m_0$. Frequently,
stable values of functions cannot be captured with absolute precision
by the calculator because the calculator holds only finitely many digits, and these may not be sufficient to express the stable value precisely. It can happen in such a case that the last digit of the successive
iteration will not settle down but rather cycle between two or more
values. Thus the test, $x = m_0$, will never be satisfied even though the
calculator has gotten as close as it is going to get to the stable value.

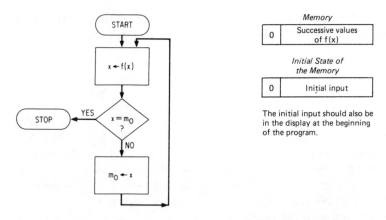

Fig. 1-15   Flowchart for stable state program (second revision)

The way to get around this difficulty is to store some very small number, say 10^{-8}, in m_1 and replace the test, $x = m_0$, by the test, $|x - m_0| < m_1$. Now the program will stop as soon as the successive values agree in the first eight digits.

Finally, we need to consider what to do in other cases where the iterations fail to settle down toward a stable value, since without adjustment of some kind, the program will run forever. To prevent this, you should start in the PAUSE format—an easy procedure if the program just outlined is written properly. Merely replace the test step and the step immediately following it with PAUSE and a GTO (or RST) order sending control back to the top of the program. Now the program will work as it did in the previous flowchart. You have here the best of both worlds, a program that will do the search automatically but that can easily be modified to provide more feedback when needed. [*Solution was realized on an HP-25 in 10 steps—not including a subprogram for computing values of f(x).*]

Notes: A potential ambiguity exists in the flowchart of Fig. 1-15. Look at the steps, $x = m_0$ and $m_0 \leftarrow x$. The second x is intended to be the same as the first, even though it may no longer be in the display after the test has been run. (See Appendix A for further discussion of how to read flowcharts.) Also notice that part of the input for this program is another program.

The program given here can sometimes be used to solve equations. Suppose that you want to solve the equation: $x^2 - x - 1 = 0$. If the x is moved to the other side of the equals sign, the equation becomes: $x^2 - 1 = x$. To find an x such that $x = x^2 - 1$ is the same as finding a stable value for the expression: $x^2 - 1$. (Think about it.) If we try the routine given above on the expression $x^2 - 1$, and if it finds a stable value, the original equation is solved. Go ahead and try it. You will find that it doesn't work. No matter where you start, the iterations never settle down. Eventually, they may start flipping back and forth between 0 and -1, even for the stable values $(1 \pm \sqrt{5})/2$. The reason is that the calculator cannot express either of these numbers precisely. As a result, the numbers you wind up with on the calculator are not really stable. All of this illustrates that even if stable values exist, you may not be able to find them by iteration.

If you manipulate the original equation another way, you *can* solve it by finding a stable value. Isolating the x^2 on the left, you get the equation: $x^2 = x + 1$. If square roots are taken on both sides, this becomes: $x = \pm \sqrt{x + 1}$. Thus the solutions of the equation are the stable values of the expressions, $\sqrt{x + 1}$ and $-\sqrt{x + 1}$. Since we have already found the stable value of the former expression, the equation is solved.

Stable values are sometimes called *fixed points*. They are one of the BIG IDEAS in mathematics. The idea of *iteration* is another one.

We have assumed throughout this section that the only type of output your calculator will produce is numbers. However, a new generation of calculators possess *alphanumeric* capabilities; that is, the calculators can output letters and words. Such calculators have a much more flexible output (and input) potential than their predecessors.

1.7 Subroutines and Flags

Difficulty: 2

Your calculator may or may not have subroutines or flags. We will now show you how to use them if you have them and how to build them into your programs artificially if you don't.

Many programs use the same sequence of key strokes several times. When they do, you might well wish that the whole sequence were just one key on your calculator, but of course it isn't. It is on such occasions that subroutines come in. A subroutine is a mini-program that can be used by the main program several times.

Consider the following problem. A spherical balloon filled with water has a radius of r_0 cm. How much water must be added to make the radius equal r_1 cm? Let us write a program that accepts as input the two numbers, r_0 and r_1, and outputs the volume of water to be added. Since the volume of a sphere of radius r is given by the equation, $V = 4/3\ (\pi r^3)$, the balloon initially has a volume of $4/3\ (\pi r_0^3)$. Since we want the resulting volume to be $4/3\ (\pi r_1^3)$, the amount to be added is the difference, $4/3\ (\pi r_1^3) - 4/3\ (\pi r_0^3)$.

The sample programs below assume that r_1 is stored in memory 1 and r_0 in memory 2:

HP	TI*	HP	TI
RCL 1	RCL 1	RCL 2	RCL 2
3	y^x	3	y^x
y^x	3	y^x	3
π	×	π	×
×	π	×	π
4	×	4	×
×	4	×	4
3	÷	3	÷
÷	3	÷	3
STO 0	=	STO − 0	=
RCL 2	STO 0	RCL 0	INV SUM 0
			RCL 0

* On a TI 58/59, the RCL and STO steps each consists of two steps.

Each of the above programs has a repeating block of steps, indicated by the braces. If your calculator has subroutines, the repeating block need *not* be written twice. Here is how we avoid doing so. The repeating block is written only once, somewhere else in the program memory, followed by $\boxed{\text{RTN}}$ (HP) or $\boxed{\text{INV}}$ $\boxed{\text{SBR}}$ (TI). Then whenever the block is needed, it is *called* as a subroutine (that is, execution is transferred to the block by a subroutine order—$\boxed{\text{GSB}}$ or $\boxed{\text{SBR}}$). When the calculator gets to the $\boxed{\text{RTN}}$ or $\boxed{\text{INV}}$ $\boxed{\text{SBR}}$ at the end of the block, it automatically transfers execution back to the step following the one that called the subroutine.

There are two methods for transferring (or *addressing*) program execution to another portion of the program memory: *direct addressing* and *relative addressing*. For some calculators (like the HP 33E) that have direct addressing, execution can be transferred directly to a certain step number. Since these calculators do not have labels, the instruction GSB 25, for instance, is used to transfer execution to step 25 of a program. For other calculators (like the HP 67/97 and TI 57) that use relative addressing,[25] a step number cannot be addressed directly and a label must be used: The step LBL 3 is placed in step 25. LBLs do not *do* anything when encountered in program execution; they merely serve as markers. To transfer execution to step 25, you put in the instruction GSB 3 (or GTO 3). When the program encounters the GSB 3, it searches for the LBL 3, finds it at step 25, and continues execution from there.

---

[25] Still others (like TI 58/59) have both capabilities.

The previous programs, rewritten using subroutines, are shown below (note that we have saved three or four program steps by using a subroutine):

HP	TI*	HP	TI
RCL 1	RCL 1	LBL 0	LBL 0
GSB 0	SBR0	3	y^x
STO 0	STO 0	y^x	3
RCL 2	RCL 2	π	x
GSB 0	SBR0	x	π
STO −0	INV SUM 0	4	x
RCL 0	RCL 0	x	4
R/S	R/S	3	÷
		÷	3
		RTN	=
			INV SBR

* On a TI 58/59, the RCL and STO steps each consist of two steps.

A subroutine can itself call another subroutine. In some calculators, the second subroutine can in turn call a third subroutine, and so on. The number of subroutine "levels" is limited by the kind of calculator you have; consult your manual to see how many are allowed.

There is an easier way to program the balloon problem—for which no subroutines are required—by using the fact that $4/3\ (\pi r_1^3)\ -\ 4/3\ (\pi r_0^3)\ =\ 4/3\ \pi\ (r_1^3\ -\ r_0^3)$:

HP	TI*
RCL 1	(
3	RCL 1
y^x	y^x
RCL 0	3
3	−
y^x	RCL 0
−	y^x
4	3
x	)
π	x
x	4
3	x
÷	π
	÷
	3

* On a TI 58/59, the RCL and STO steps each consist of two steps.

What if your calculator does not allow subroutines? The repeated block can still be set aside somewhere in the program memory and called by a GTO statement when needed by the main program. The problem is, how do we get the program pointer back to the correct place in the main program? Here is a way.

We will use a memory (m_3 in the flowchart of Fig. 1-16) to keep track of the progress of the main program. Before we call the "subroutine," we store a number in m_3 to tell us where we are in the main program. At the end of the "subroutine," we check the number in m_3 to find out where to go in the main program. The extra register m_3 contains a 1 when the "subroutine" is first called and a 0 the second time. The test, $m_3 \neq 0$, thus tells the calculator where to re-enter the main program.

It seems a waste to use an entire 10 to 13-digit memory merely to store either 0 or 1. This is the reason that some calculators employ flags. A flag is actually a memory that can be in one of two states—set (=1) or clear (=0). A flag can be set, cleared, or tested. As an example, the program of Fig. 1-16 is shown rewritten with a flag—flag 0—in Fig. 1-17.

If flags are not available, you can use a memory as a flag, just as we previously used m_3. But what if you have no memories to spare for this purpose? Well, here are two tricks.

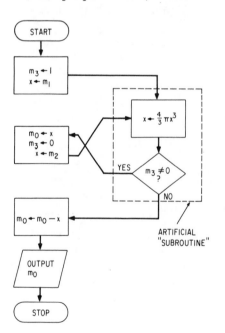

Memory	
0	$4/3 \, \pi \, r_1{}^3$
1	r_1
2	r_0
3	0 or 1

Fig. 1-16 Flowchart to keep track of progress of balloon program

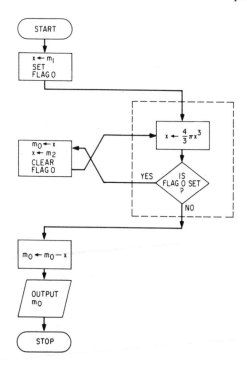

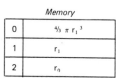

Memory	
0	$^{4}/_{3} \pi r_1{}^3$
1	r_1
2	r_0

Fig. 1-17 Flowchart for balloon program rewritten with a flag

First, if one of the memories in your program, say m_0, stores only positive numbers, it can double as a flag. To clear this "flag," use the following: $m_0 \leftarrow |m_0|$. To set the "flag", use the following: $m_0 \leftarrow -|m_0|$. Thus, the "flag" is cleared if $m_0 > 0$ and set if $m_0 < 0$. To test the "flag", use the test: $m_0 < 0$. Take warning, however. If you need the contents of m_0 in the program, remember that m_0 is supposed to be positive and that each RCL 0 must therefore be followed by an ABS (HP) or an $|x|$ (TI) statement.

Second, if your calculator has two angular modes (degrees and radians) and if no trigonometric functions are used in the program, the angular mode can be used as a flag. To clear this "flag," put the calculator in degree mode; to set it, put the calculator in radian mode. How is the "flag" tested? Note that since

$$\cos \pi = \begin{cases} .99849715 \ldots & >0 \text{ in degree mode} \\ -1 & <0 \text{ in radian mode} \end{cases}$$

the test, $\cos \pi < 0$, will test the flag.

Problem: Write a program that accepts as input numbers a and b and calculates

$$c = \frac{\sqrt{a^2 + 3} - a}{\sqrt{b^2 + 3} - b}$$

Solution: The numerator and denominator of c are the same, except for the interchange of a and b. We will thus write a subroutine that will take the number x in display and calculate $\sqrt{x^2 + 3} - x$. We call this subroutine twice—once with x = a, for the numerator, and once with x = b, for the denominator.

The flowchart for this program (Fig. 1-18) begins with a in m_0 and b in m_1. It actually calculates $(a - \sqrt{a^2 + 3})/(b - \sqrt{b^2 + 3})$, but this expression equals $(\sqrt{a^2 + 3} - a)/(\sqrt{b^2 + 3} - b)$, or c.

1.8   Loops and Nested Loops

Difficulty: 2

In many of the problems in this book, you will find that a sequence of program steps must be executed several times *in a row*. A simple way to do this would be to repeat the same block of steps in the program memory as many times as required. But if this block of steps is too long, or if many repetitions are required, you are needlessly wasting precious programming space. To avoid such waste, you must use what is called a "loop." You write the block of steps only once in the program memory, and you place a GTO statement at the end of the block to return execution to the beginning of the block.

Problem 1:   Write a program that calculates, and pauses to output, the numbers, cos (1), cos (2), cos (3), and so on.

This problem requires what is called an *endless loop*. In theory, the calculator will just keep going on, forever calculating cosines. Actually, after 10^{10} to 10^{13} iterations (depending on the accuracy of the calculator), the loop will start to calculate the cosine of the same large number over and over again because the 1 that the calculator attempts to add on just gets rounded off. Don't hold your breath, however. The HP-25 can do about 38 iterations per minute and consequently won't complete 10^{10} iterations for about 500 years!

Much more common are loops programmed to terminate after a fixed number, k, of iterations. For these, you need an extra memory, called a *counter*, to keep track of the number of iterations done so far, and also a test to terminate the program when k iterations of the loop have been executed. Two methods are available: *count-down* and *count-up*. In the flowcharts of Figs. 1-19(a) and 1-19(b), B denotes the block of steps to be repeated k times.

The count-down loop is usually more convenient, for two reasons. First of all, in the count-up loop, since the number k is used for

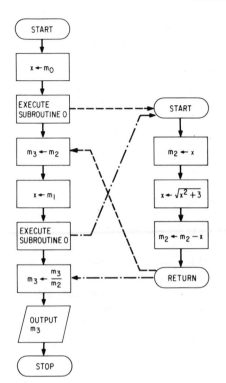

Memory	
0	a
1	b
2	$x-\sqrt{x^2+3}$
3	$a-\sqrt{a^2+3}$

Fig. 1-18 Flowchart for a program using a substitute "flag"

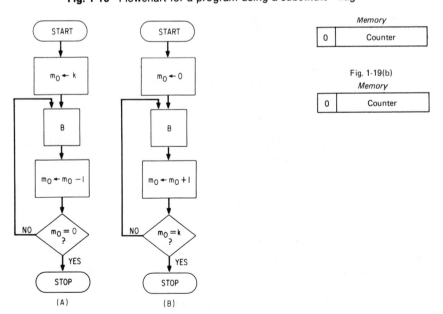

Fig. 1-19 Flowcharts for (a) count-down and (b) count-up loops

each run through the loop, it must either be entered directly in the program or, more often, be stored in another memory. In the count-down loop, however, the number k is used only once—at the beginning. Second, if you use the count-down loop, the DSZ instruction [see Sec. 1.2(5) for a full discussion] can be conveniently used both to decrement the counter and to test whether k iterations have been performed—all in one fell swoop.

A small difficulty sometimes occurs in iterating a loop k times. If $k = 0$, you want the calculator to skip the loop entirely. Notice, however, that in both flowcharts of Figs. 1-19(a) and (b), the input, $k = 0$, causes an endless loop. (In the count-down loop, for example, the first time the test, $m_0 = 0$?, is executed, $m_0 = -1$. The test fails, therefore, and the loop is executed again. The values of m_0 get successively smaller than -1 and never pass the test, $m_0 = 0$?)

One way to handle this difficulty is to place the test at the beginning of the loop, instead of at the end. Two variations of the count-down loop are shown in Figs. 1-20(a) and (b).

If you have a DSZ key, it can't be used in the flowchart of Fig. 1-20(a) as it stands; the flowchart of Fig. 1-20(b) is therefore better. But if you don't have a DSZ, you might as well use the flowchart of Fig. 1-20(a) to save the trouble of adding 1 to k at the beginning.

Problem 2:   Write programs that accept as input a positive integer n and output the quantity,

$$1^2 + 2^2 + 3^2 + \ldots + (k - 1)^2 + k^2$$

using

(a) A count-down loop with a test at the end
(b) A count-up loop with a test at the end
(c) A count-down loop with a test at the beginning
(d) A count-up loop with a test at the beginning

Before you proceed, let us introduce you to a very useful piece of mathematical shorthand. The Greek letter Σ (sigma) is used to represent in abbreviated form the sum of a lengthy series of numbers. The sum, $1^2 + 2^2 + 3^2 + \ldots + (k - 1)^2 + k^2$, for example, is written as $\sum_{i=1}^{k} i^2$.

The formal definition of such an expression is as follows: If a and b are integers, with $a \leq b$, and if f is a function that assigns to each integer i in the range $a \leq i \leq b$ some number f(i), then the expression, $\sum_{i=a}^{b} f(i)$, means that we must take all integers i in the range

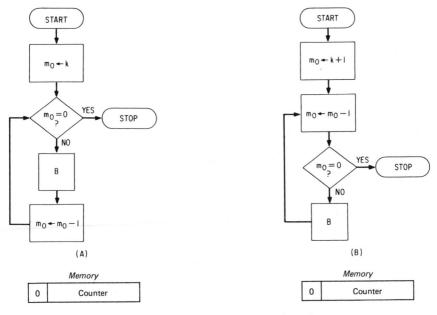

Fig. 1-20 Two variations of the count-down loop

$a \leq i \leq b$, apply the function f to each one, and add up the results. In problem 2, therefore, $a = 1$, $b = k$, and $f(i) = i^2$.

Here are some more examples:

$$\sum_{i=-3}^{0} (i^2 - 1) = [(-3)^2 - 1] + [(-2)^2 - 1] + [(-1)^2 - 1] + [(0)^2 - 1] = 10$$

$$\sum_{s=0}^{3} \frac{1}{s+1} = \frac{1}{0+1} + \frac{1}{1+1} + \frac{1}{2+1} + \frac{1}{3+1} = 2\frac{1}{12}$$

$$\sum_{k=3}^{7} \text{INT}(\sqrt{k}) = \text{INT}(\sqrt{3}) + \text{INT}(\sqrt{4}) + \text{INT}(\sqrt{5}) + \text{INT}(\sqrt{6}) + \text{INT}(\sqrt{7})$$

$$= 1 + 2 + 2 + 2 + 2 = 9$$

$$\sum_{i=15}^{15} (i - 3) = (15 - 3) = 12$$

It is not apparent how many times some loops are to be executed. Since these loops contain a test, they are not endless, but there are many possibilities for the test. Perhaps the loop is to be executed until two variables are equal, for example, or until their difference is less than some preassigned value, or until one is less than the other. Here is one such loop for you to try.

Problem 3:   Write a program that accepts as input a pair of integers a and b (possible negative) with a $\leq$ b and that outputs

$$\sum_{i=a}^{b} i^4$$

A more difficult problem would be to find a solution, x, to the equation: x = cos x. One way to do so is as follows: Start out with any number x_0 as an initial guess. Generate an infinite sequence of numbers—x_1, x_2, x_3, x_4, . . .—by the rule: $x_1 = \cos x_0$, $x_2 = \cos x_1$, $x_3 = \cos x_2$, $x_4 = \cos x_3$, ... As each new x_i is generated, test to see whether or not $x_i = x_{i-1}$. If not, generate the next number and keep going. Otherwise, if $x_i = x_{i-1}$, we have $x_i = \cos x_{i-1} = \cos x_i$. Thus, $x = x_i$ is a solution, and the loop terminates.

Problem 4:   Write a program to find a solution, x, to the equation,

$$x = \cos x$$

using the technique described above, with $x_0 = 0$. Run your program in both degree and radian mode (the answers will be different).

Sometimes a program will require two or more loops, one "nested" inside the other. Referring back to the flowchart in Fig. 1-19(a), imagine that block B contains a block C that is repeated e times by another loop, as shown in the flowchart of Fig. 1-21.

Before we present a problem of this sort, let us introduce some more mathematical shorthand, this time using the Greek letter Π (pi). If a and b are integers, with a $\leq$ b, and if f is a function that assigns to each integer i in the range a $\leq$ i $\leq$ b a number f(i), then the expression,

$$\prod_{i=a}^{b}$$

means that we must take all integers i in the range a $\leq$ b, apply f to each one, and *multiply* the answers. For example, if a = −3, b = 1, and f(i) = i^2 + 1, then

$$\prod_{i=-3}^{1} (i^2 + 1) = [(-3)^2 + 1]\, [(-2)^2 + 1]\, [(-1)^2 + 1]\, [(0)^2 + 1]\, [1^2 + 1]$$

$$= (10)\, (5)\, (2)\, (1)\, (2) = 200$$

Problem 5:   Write a program, using the flowchart in Fig. 1-21, that accepts as input any pair, k and e, of positive integers and outputs the number,

$$\sum_{i=1}^{k} \left[\prod_{j=1}^{e} (i + j) \right]$$

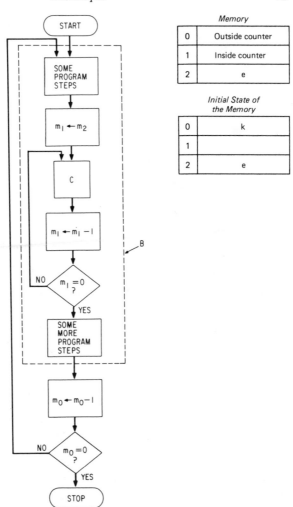

Fig. 1-21 Flowchart of a nested loop

Solutions to Problems

Solution to Problem 1: The flowchart in Fig. 1-22 tells it all. We use memory m_0 to store successively the numbers 1, 2, 3, ..., whose cosine we desire.

Solution to Problem 2: To solve Problems 2(a) and 2(b), we will of course use the flowcharts of Fig. 1-19(a) and (b), respectively. To solve Problem 2(c), we can use the flowchart of either Fig. 1-20(a) or (b). We will tell you only the contents for the mysterious block B in the four flowcharts.

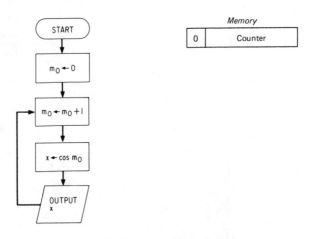

Fig. 1-22   Flowchart for Problem 1

We will use memory m_1 to accumulate the answer. If the program is counting down, m_1 will successively contain the following numbers:

$$k^2, (k - 1)^2 + k^2, (k - 2)^2 + (k - 1)^2 + k^2, \ldots$$

while if the program is counting up, m_1 will successively contain these numbers:

$$1^2, 1^2 + 2^2, 1^2 + 2^2 + 3^2, \ldots$$

Notice that the number to be added to m_1 is precisely the square of the counter value, m_0. The contents for block B are thus as shown in Fig. 1-23.

There is just one thing to add: The step, $m_1 \leftarrow 0$, should be placed at the beginning of each of the flowcharts in Figs. 1-19 and 1-20 to prepare the accumulator m_1. In the count-up loop of Problem 2(b), k must also be stored in some memory, say m_2. The flowchart in Fig. 1-24 shows one way to solve Problem 2(d).

Solution to Problem 3:   The loop counter m_0 will start at b and work its way down to a. As in Problem 2, we will accumulate the sum in register m_1. See the flowchart in Fig. 1-25.

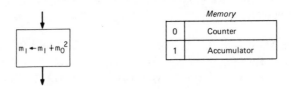

Fig. 1-23   Contents of block B

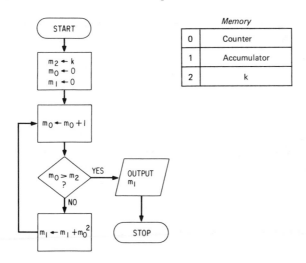

Fig. 1-24 Flowchart for Problem 2(d)

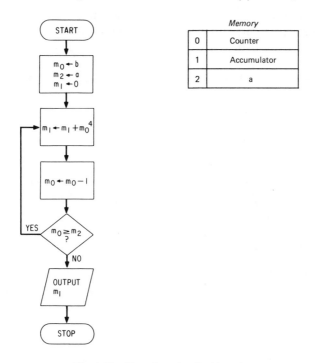

Fig. 1-25 Flowchart for Problem 3

Solution to Problem 4: You don't need a counter here, although you could put one in if you want to know the number of iterations the program requires. You need only one memory, m_0, to store the amount x_i. The program begins with x_0 in the x register. See the flowchart in

Fig. 1-26. [Solution in degree mode, x = .998477415, was realized on an HP67 in five iterations; solution in radian mode, x = .7390851332, was realized in 59 iterations. The number of iterations taken varies from calculator to calculator, depending on their accuracy, but your final answers should be approximately the same as ours.]

Solution to Problem 5: Using the flowchart in Fig. 1-21, we set up an "outside" loop with its counter $m_0 = i$, starting at $i = k$ and counting down. Within this loop, we set up an "inside" loop, with its counter $m_1 = j$, starting at $j = e$ and counting down. The inside loop calculates

$$i + j = m_0 + m_1$$

and accumulates in register m_3 the product

$$\prod_{j=1}^{e} (i + j)$$

The outside loop accumulates the total sum in register m_2. We need one more memory, m_4, to store input e. See the flowchart in Fig. 1-27.

 To test your program, try $k = 4$ and $e = 3$. Then,

$$\sum_{i=1}^{4} \prod_{j=1}^{3} (i + j) = \prod_{j=1}^{3} (i + j) + \prod_{j=1}^{3} (2 + j) + \prod_{j=1}^{3} (3 + j) + \prod_{j=1}^{3} (4 + j)$$

$$= 2 \cdot 3 \cdot 4 + 3 \cdot 4 \cdot 5 + 4 \cdot 5 \cdot 6 + 5 \cdot 6 \cdot 7$$

$$= 24 + 60 + 120 + 210 = 414.$$

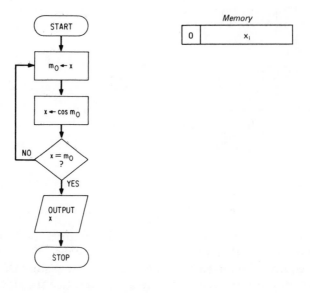

Fig. 1-26 Flowchart for Problem 4

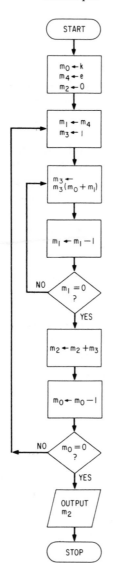

Memory	
0	Outside counter
1	Inside counter
2	Σ accumulator
3	Π accumulator
4	e

Fig. 1-27 Flowchart for Problem 5

Notes: See Sec. 1.6 for a discussion of Problem 4 from another point of view.

1.9 Indirect Addressing

Difficulty: 3

In this section, we will show you how to use *indirect addressing,* if your calculator has this feature, and how to "fake it" if it does not.

Suppose that there are four numbers stored in memories 1 through 4 of your calculator. The instructions RCL 1, RCL 2, RCL 3, and RCL 4 will of course bring the numbers m_1, m_2, m_3, and m_4, respectively, into the display. In certain situations, however, another part of your program will calculate a number j (where j = 1, 2, 3, or 4), and you may want the program to bring the corresponding number m_j into the display.

Problem 1:   Write a program that will accept as input a number j (where j = 1, 2, 3, or 4) and output the contents of memory j. If you solve this problem, you may notice that it takes a lot of programming steps to do a rather simple thing. There ought to be an easier way! If your calculator has indirect addressing, there is. Since different calculators have different instructions for using indirect addressing, however, let's take a quick look at some typical examples.

In the HP 19C/29C, the indirect addressing instruction RCL i is executed as follows: The calculator examines the contents of the special indirect register m_0. Then it goes to the register whose address is in m_0 and recalls the contents of that register. For example, if $m_0 = 5$, the instruction RCL i is the same as the instruction RCL 5. Moreover, the instructions STO i, STO + i, STO − i, STO × i, and STO ÷ i are the same as the instructions STO 5, STO + 5, STO − 5, STO × 5, and STO ÷ 5, respectively (if $m_0 = 5$, of course). Thus, the following sequence of steps will solve problem 1 if j is in the display to begin with:

STO 0
RCL i

The HP 67/97 works similarly, except that the special indirect register is memory I rather than memory 0. The corresponding steps for Problem 1 now become

STO I
RCL i

The TI 58/59 does not have a special indirect register; any memory may be used. For example, if memory 7 contains the number 3, steps RCL IND 7 and STO IND 7 are identical, respectively, to steps RCL 3 and STO 3. Any memory, not just m_7, can be used as the indirect register. We could use m_5 to solve Problem 1 as follows:

STO 5
RCL IND 5

(Of course, if your calculator does not have indirect addressing, you are stuck with faking it, as we do in our solution to Problem 1.)

The instruction DSZ [see Sec. 1.2(5)] is often used in conjunction with indirect addressing when the contents of a large block of registers must be manipulated, the reason being that the DSZ key can be used to decrement the indirect register.

The following problem assumes that your calculator has indirect addressing.

Problem 2: Write a program that outputs the sum of the squares of registers m_1 through m_6.

Solution to Problem 1: We want our program to branch to one of the four steps, RCL 1, RCL 2, RCL 3, or RCL 4. We begin with j in the x register and successively subtract 1 from x until x = 0. See the flowchart in Fig. 1-28.

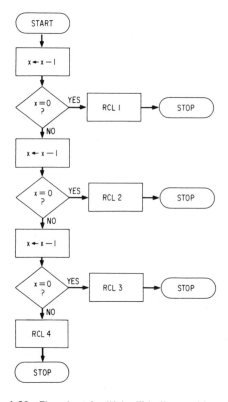

Fig. 1-28 Flowchart for "faked" indirect addressing

Solution to Problem 2:   We set up a loop that successively recalls memories m_6 through m_1, squares the contents, and accumulates the sum in register m_7. Following is the program for the HP 19C/29C (we leave you the task of translating it to whatever calculator you have):

<table>
<tr><td>*LBL 0</td><td></td><td></td></tr>
<tr><td>6</td><td colspan="2">*Memories*</td></tr>
<tr><td>STO 0</td><td>0</td><td>Indirect addresser</td></tr>
<tr><td>0</td><td>7</td><td>Sum accumulator</td></tr>
<tr><td>STO 7</td><td></td><td></td></tr>
<tr><td>*LBL 1</td><td></td><td></td></tr>
<tr><td>RCL i</td><td></td><td></td></tr>
<tr><td>x^2</td><td></td><td></td></tr>
<tr><td>STO + 7</td><td></td><td></td></tr>
<tr><td>DSZ</td><td></td><td></td></tr>
<tr><td>GTO 1</td><td></td><td></td></tr>
<tr><td>RCL 7</td><td></td><td></td></tr>
<tr><td>RTN</td><td></td><td></td></tr>
</table>

Note: If your calculator does not have indirect addressing, you can still solve Problem 2 by manufacturing an indirect addresser as described in the solution to Problem 1. You will find, however, that it takes a lot of program space and illustrates why indirect addressing is such a nice feature to have built in.

1.10   Computing y^x Accurately

Difficulty: 3

Suppose that you want to compute 2^{26}. You will probably use your y^x key to save yourself the trouble of multiplying 2 by itself 26 times. If you do, you may not get the right answer. Try it on your calculator. Even if you get the correct answer (67,108,864), your calculator might be fooling you. Take the fractional part; it should be zero since 2^{26} is obviously a whole number.

The reason that your calculator may fail to get the right answer is that it uses logarithms to make the computation. The natural log of 2^{26} is $\ln(2^{26})$, or $26 \ln(2)$. Compute the latter number. This yields the natural log of the number you want to find, and you can get the number itself by pressing e^x (INV lnx on some calculators). You should then get the same answer that your calculator gave when you used the y^x key to find 2^{26} (an exception is the HP 67, which uses logs to find y^x if x is not a whole number but manages to get the precise answer if it is).[26] The advantage of using logarithms to compute y^x is

---

[26] For a more thorough discussion of the use of logarithms in computing exponential expressions and large products, see Sec. 1.2(8).

that the technique works whether or not x is a whole number. The disadvantage, as we have seen, is that the answer is not exact.

Let us now examine another problem. We want to write a program for computing y^x precisely when x and y are positive whole numbers. We will consider the problem only for those numbers x and y for which y^x is small enough for your calculator to hold all the digits. For larger numbers, there is no way to get the answer precisely.[27] Assuming that your calculator will display ten digits, we would like a program that will yield the precise value of y^x whenever $y^x < 10^{10}$ (the first eleven-digit number). One solution is to have the calculator multiply y by itself x times, but this algorithm is too slow (see Sec. 1.16). There is, however, another possibility. Think about the problem a bit before reading the next paragraph.

The first approach is to compute y^x using the y^x key and then round off the result to the nearest whole number. If x and y are positive whole numbers, y^x should clearly be a whole number. Then if the calculator's value for y^x is close enough, rounding to the nearest whole number will get it exactly. But this method raises two questions: (1) Just how do you get the calculator to round off to the nearest whole number? (2) Does the process outlined always work (that is, does it always produce y^x precisely)?

The answer to the first question is not too difficult. There is an elegant solution requiring five steps at the most. If you have a TI 58/59 or HP 56/97, the problem is even easier (see the Notes). We leave it to you to figure this out. The second question is the real heart of the matter. The answer here is not so simple. In fact, the solution works on some calculators but not on others. You are going to have to investigate your own calculator to see whether the process outlined always works or not. This, in fact, is the problem we want you to solve.

Problem: Construct a "diagnostic program" that will tell you whether or not computing y^x with the y^x key and rounding off to the nearest whole number will always give you the correct value of y^x when x and y are positive whole numbers and $y^x < 10^{10}$. Your program should generate all possible values of y^x smaller than 10^{10} using the method outlined above and should compare these with the precise values (which the program will also have to generate). If the method fails for some pair of numbers x and y, the program should halt so that you can see what x and y are. You will then need to do further analysis to see exactly where the method breaks down so that it may be modified and made failproof. Whatever the outcome, this program is going to take a *long time* to execute. Happy hunting!

---

[27] Unless you use double-precision arithmetic.

Solution: First, let's dispose of the first question. In order to round a (positive) number off to the nearest whole number, just add 0.5 and take the integer part. Think about it and you will see that it works. Give yourself an A if you got this solution.[28]

Now for the second, and principal, question. What we need is a systematic way of generating all pairs of positive whole numbers x and y such that y^x is less than 10^{10}. As each pair is generated, we will also want to generate the precise value of y^x for comparison with the value obtained using the y^x key and rounding off to the nearest whole number.

The routine will go basically as follows: For each y we will start at y^2 and successively build all powers of y (y^3, y^4, y^5, . . .) until an exponent x is reached such that $y^x > 10^{10}$. At this point, we go back to the beginning, increase y by 1, set x to 2, and start over. We continue until the first power of y (that is, y^2) is greater than 10^{10}, at which point we shall terminate the whole routine. Each successive power of y will be built from the previous power by multiplying it by y. These will be the precise values of y^x that will then be tested against the value obtained using the y^x key and rounding off. If the test ever fails, the program will stop.

Here are the details. Four memories will be used: y will be stored in m_0, x in m_1, y^x in m_2, and 10^{10} in m_3. The program will contain two "nested loops" (see Sec. 1.8). The inner loop will generate a new value of x and test the pair y^x. The outer loop will use enough inner loops to generate powers of a given value of y until a power exceeding 10^{10} is reached. Then the loop will set y and x back to 2 and start over.

We will describe the program by starting at the top of the outer loop, which will also be the beginning of the program. (You may want to follow the flowchart in Fig. 1-29 as you read this.) Increase the contents of m_0 by 1 (since m_0 has a 1 in it when the program starts, this will set m_0 at 2, which is what it should be). Put a 1 in m_1 and put the contents of m_0 in m_2 (this sets m_2 at y^1).

Now check to see if $m_0^2 > 10^{10}$. If so, you have generated all the pairs and can now terminate the program. If not, pass to the inner loop. Add 1 to the contents of m_1 (if this is the first time into the loop, m_1 will now be 2, which is what it should be). Multiply the contents of m_2 by the contents of m_0, leaving the result in m_2. You have now increased x by 1, and you have the precise value of y to the new x power in m_2.

Check to see if $m_2 > 10^{10}$ (the contents of m_3). If so, you have generated all the powers of this particular y that you need. Loop back to the top of the outer loop and again increase the contents of m_0 by

---

[28] For further discussion, see Sec. 1.12.

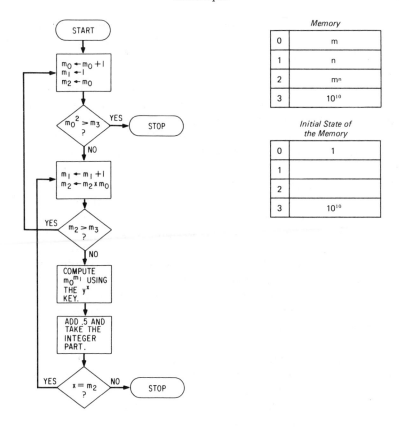

Fig. 1-29 Flowchart for accurate computation of y^x

1, etc. If not, compute m_0 to the m_1 power using the y^x key and round off to the nearest whole number.

Now check to see if the result is equal to the contents of m_2. If so, the method has worked for this particular x and y. Loop back to the ïnner loop and again multiply the contents of m_2 by the contents of m_0, etc. If not, the method has failed and the program should be stopped. This completes the program. [*Solution was realized on an HP-25 in 24 steps.*]

This program might stop for two reasons. First, the method might have worked for all x and y. If so, the program will stop because a value of y has been reached such that $y^2 > 10^{10}$. One of these two numbers will then be in the display. Or second, if the method failed for some pair x and y, then the program will stop with a number less than 10^{10} in the display (either a correct or incorrect value of y^x). *Warning:* If the program runs through all pairs without finding an error, it will operate on the order of 100 hours before stopping.

If the method has failed, you will want to determine the pairs for which it is failing. Recall the contents of m_0 and m_1. Record these along with the correct (m_2) and incorrect values of y^x. You can now continue the search for more x's and y's by returning the program to the inner loop and multiplying the contents of m_2 by the contents of m_0, etc. (see the second sentence three paragraphs back) and starting it up again. What you will find is that the method breaks down for those x's and y's that produce a y^x of nine or ten digits—a result that can be expected on calculators without guard digits.

A calculator display that shows ten digits may or may not have hidden digits in reserve. The HP-25, for example, shows everything it has, whereas the TI-57 shows 10 digits but actually holds 13. The three hidden digits, called *guard digits,* are used to improve the accuracy of the calculator's internal computations. Guard digits have their advantages and disadvantages. In the present situation, they are an advantage because they allow the calculator to compute y^x accurately enough for you to get its correct whole number value by rounding off. Guard digits become an annoyance should you want to know the exact number the calculator is holding. Although there are ways of digging out the hidden digits, the calculator is incapable of showing all of them at once.

If your calculator passed the diagnostic test, producing the actual program for computing y^x for whole numbers x and y will be easy. Just use the y^x key and round off to the nearest whole number. If this method doesn't work for large values of y^x, you will have to be a little more devious.

The trick is to break y^x into two parts small enough to be computed precisely by our method and then multiply them to get the precise answer. Suppose, for example, that you want to calculate 2^{29}. First, you must compute the values of 2^{15} and 2^{14} by rounding and then multiply them to get 2^{29}. The only real problem lies in generating the 15 and the 14.

Let us say that you have x in register m_0 and y in m_1 (you may be able to handle things more efficiently with x and y somewhere else; see Sec. 1.5). Recall x into the display and divide by 2. Since the result will not be a whole number if x is odd, take the integer part to turn it into a whole number. Keeping this number in the display, subtract it from the contents of m_0 using register arithmetic. Now you have two whole numbers, one in the display and one in m_0, whose sum is x (although x has now disappeared, don't worry); each of these numbers approximately equals x/2. Using the y in m_1 and the x in the display (rounding off, of course), now compute y^x. Compute it again with the same y but with the x in m_0. Multiply the two values together and you're done. The flowchart for this method is shown in Fig. 1-30. [*Solution was realized on an HP-25 in 21 steps.*]

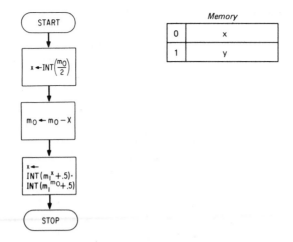

Fig. 1-30 Flowchart for accurate computation of y^y where y^x is broken into two pieces

Notes: Some calculators have their own devices for rounding off numbers in the display, for example, the TI-58/59 and the HP-67/97. This can be quite a useful function. In the present situation, you would set the display to zero decimal places and apply the rounding function to get the nearest whole number. The HP-67/97 have a special key for this function; the TI-58/59 use the sequence $\boxed{\text{EE}}$, $\boxed{\text{INV}}$, $\boxed{\text{EE}}$ to accomplish the same thing.

There are times when you will want to compute y^x precisely for an x that is a negative whole number. For example, in multiple storage, you may want to multiply a number by 10^{-r} in order to store it in the rth decimal place of a number. To do so, you first find 10^r and then take its reciprocal. In the game, Number Jotto (see Sec. 3.17), however, another trick is used, in which 10^{-r} is rounded off to an accurate value by adding it to a number with a 1 to the left of the decimal point, thereby making the incorrect digits at the end of the number fall off (see Sec. 1.14 for a more thorough discussion).

1.11   Randomizers

Difficulty: 2

Many of our programs require a random number generator, that is, a routine that will produce a random number or a sequence of random numbers. The main problem here, of course, is that no calculating device, from the programmable calculator to the largest computer, can operate truly "randomly." Turn your calculator on, punch any sequence of keys, and observe the results. Turn it on again tomorrow (or next year), punch the *same* sequence of keys, and you will get the *same* results. Computer scientists and mathematicians have attempted to get around this problem by developing the so-called *pseudo-random number* generator. Many sophisticated pseudo-random number generators have been found, but for our applications we don't need to get very fancy (in fact, we can't, since there is seldom enough programming room).

Let us begin by describing some fundamental pseudo-random number generators that are easy to implement, ones that will generate a sequence of random numbers x in the range, $0 \leqslant x < 1$. Each number x thus generated will play two roles. First, they are the required random numbers, of course. Second, each number is used as fuel to generate the next one. At the heart of these routines is some function f, which, given any number x in the range, $0 \leqslant x < 1$, returns a new number, f(x), in the same range, that is, $0 \leqslant f(x) < 1$. To use these generators, you begin by storing an arbitrarily chosen number x_0 ($0 \leqslant x_0 < 1$) in some memory, say m_0. The program outputs the first random number, x_1, by the formula, $x_1 = f(x_0)$. It also replaces the x_0 in m_0 with the new number, x_1. The same routine is again used to generate the second pseudo-random number, x_2. Since m_0 now contains x_1, the formula becomes: $x_2 = f(x_1)$. Thus, starting from the "seed," x_0, an infinite sequence—$x_1, x_2, x_3, x_4, \ldots$ —of pseudo-random numbers is generated by the formula, $x_i = f(x_{i-1})$, where $i = 1, 2, 3, 4. \ldots$

Different formulations of the function f, of course, will produce different generators, some more "random" than others. The function, f(x) = FRAC(x + .5), for example, would be a horrible choice. If the seed, x_0, were .3125, for example, this function would generate a sequence—x_1 = .8125, x_2 = .3125, x_3 = .8125, x_4 = .3125, etc.—that is hardly random. Here are two better functions:

$$f(x) = \text{FRAC}(997x)$$
$$f(x) = \text{FRAC}\left((\pi + x)^5\right)$$

Either may be used for any randomizer in this book. Although both have drawbacks (see Notes), they are good enough for our purposes. Your owner's manual may suggest others, or you might invent your own.

Here are several problems for you to try, using any of the basic randomizers described above.

Problem 1:   Write a program that accepts as input a seed, s, in the range, $0 \leqslant s < 1$, and any positive number, c, and then outputs a sequence of random numbers, x_i, in the range, $0 \leqslant x_i < c$.

Problem 2:   Write a program that accepts as input a seed, s, in the range, $0 \leqslant s < 1$, and a pair of numbers, a and b, such that a < b, and then outputs a sequence of random numbers, x_i, in the range, $a \leq x_i < b$.

Problem 3: Write a program that accepts as input a seed, s, in the range, $0 \leq s < 1$, and a positive integer, k, and then outputs a sequence of random *integers,* x_i, in the range, $0 \leq x_i \leq k$.

Problem 4: Write a program that accepts as input a seed, s, in the range, $0 \leq s < 1$, and a pair of integers, m and n, such that $m < n$, and outputs a sequence of random integers, x_i, in the range, $m \leq x_i \leq n$.

Solutions to Problems

Solution to Problem 1: Using any of our basic randomizers to produce a random number between 0 and 1, we multiply this output by c to obtain a random number between 0 and c (see the flowchart in Fig. 1-31, in which f represents the randomizing function).

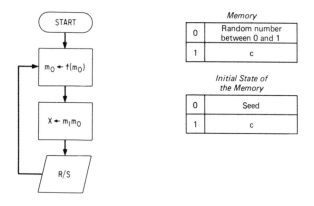

Fig. 1-31 Flowchart for outputting a sequence of random numbers in the range, $0 \leq x_i < c$

Solution to Problem 2: By using Problem 1 to generate a random number x in the range, $0 \leq x < (b - a)$, and adding a to this number, we get a random number between a and b (see the flowchart in Fig. 1-32).

Solution to Problem 3: Using our solution to Problem 1 to generate a random number, x, in the range, $0 \leq x < (k + 1)$, and taking the integer part of the output, we get a random integer between 0 and k, inclusive (see the flowchart in Fig. 1-33).

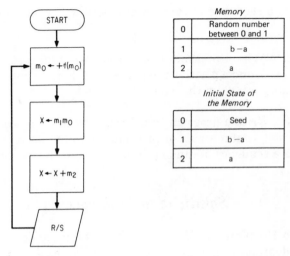

Fig. 1-32 Flowchart for outputting a sequence of random numbers in the range,
$a \leq x_i < b$

Solution to Problem 4:   Using our solution to Problem 3 to generate
a random integer, x, in the range, $0 \leq x \leq (n - m)$, and adding m to
this integer, we get a random integer between m and n (see the flowchart in Fig. 1-34).

Notes: The main drawback to our first randomizing function, $f(x) =$
FRAC(997x), is that certain seeds do not work well. For example, when

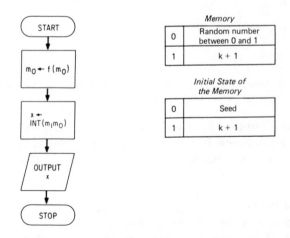

Fig. 1-33 Flowchart for outputting a sequence of random numbers in the range,
$0 \leq x_i \leq k$

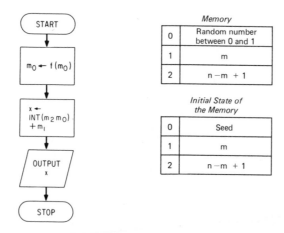

Fig. 1-34 Flowchart for outputting a sequence of random numbers in the range, $m \leq x_i \leq n$

$x_0 = 0$, the sequence produced—$x_1 = 0$, $x_2 = 0$, $x_3 = 0$, . . .—is hardly a sequence of random numbers. The HP 67/97 standard pack recommends that x_0 should be chosen so that neither $(1/2)(x_0)(10^7)$ nor $(1/5)(x_0)(10^7)$ are integers. The HP 67/97 games pack goes further to recommend that x_0 consist of the nine digits 1 through 9, following the decimal point in some order and ending in 1, 3, 7, or 9.

Our second randomizing function, $f(x) = \text{FRAC}[(\pi + x)^5]$, which we found in the HP 25 applications manual, is not nearly so sensitive to the choice of seed. Its drawback, however, is its slight tendency to produce numbers closer to 0 than to 1.

1.12 The Magic Key: INT

One of the most valuable keys on your calculator—frequently used in this book—is the $\boxed{\text{INT}}$ (Integer Part) key. We took advantage of it in the last two problems of the previous section, and it lies at the heart of two basic techniques, changing bases and multiple storage, that we will discuss in the next two sections. Since almost all problems in Chap. 3 use a randomizer, multiple storage, or both, they consequently use $\boxed{\text{INT}}$.

A few of the older programmable calculators, such as the SR-52, have no $\boxed{\text{INT}}$ key (see your owner's manual for a way to obtain this function should it not be provided).

The basic function of the $\boxed{\text{INT}}$ key is to allow you to divide a number in two at the decimal point, thereby eliminating that portion of the x register to the right of the decimal point, called the *fractional part* of x and leaving only the portion to the left of the decimal point, called the *integer part* of x. Thus, INT (3.015) = 3, INT (−371) = −371, INT (−3.015) = −3, and INT (2.301576128 × 10^{12}) = 2.301576128 × 10^{12}. (In the last example, the .301576128 is *not* eliminated because the decimal point does not actually fall after the first digit 2; in non-scientific notation, 2.301576128 × 10^{12} = 2301576128000.00.)

There are also some useful variations on the $\boxed{\text{INT}}$ key that you should know. The "greatest integer" function, for instance, is important in many applications. For any number x, the symbol $\lfloor x \rfloor$—the traditional notation is [x]—is defined to be the greatest integer less than or equal to x. Other names for $\lfloor x \rfloor$ are "x round down" and "the floor function." Thus, $\lfloor 3.015 \rfloor$ = 3, and $\lfloor 4217 \rfloor$ = 4217, and it would seem that there is no difference between INT(x) and $\lfloor x \rfloor$. In fact, INT(x) = $\lfloor x \rfloor$ only if x ≥ 0 or is already an integer. Note that INT (−3.05) = −3, whereas $\lfloor -3.05 \rfloor$ = −4 (the reason is that −4 < − 3.05 < − 3.)

Problem 1:　Write a program that will accept as input any number, x, and output $\lfloor x \rfloor$.

Closely related to the greatest integer function $\lfloor x \rfloor$ is the "least integer" function $\lceil x \rceil$, also called "x round up" or "the roof function," which is defined as the least integer greater than or equal to x. Thus, $\lceil 3.01 \rceil$ = 4, $\lceil 3 \rceil$ = 3, $\lceil -3 \rceil$ = −3, and $\lceil -3.01 \rceil$ = −3.

Problem 2:　Write a program that will accept as input a number, x, and output $\lceil x \rceil$.

If a program requires both $\lfloor x \rfloor$ and $\lceil x \rceil$, steps can be saved with a trick using subroutines (see Sec. 1.7). Set aside a label, say label 0, to execute $\lfloor x \rfloor$, as in Problem 1. When $\lfloor x \rfloor$ is required, call subroutine 0, using GSBO with the HP and SBRO with the TI. If $\lceil x \rceil$ is required, use the following steps:

HP	TI
CHS	+/−
GSBO	SBRO
CHS	+/−

(The reason these steps work is that $\lceil x \rceil$ = $-\lfloor -x \rfloor$. Try this equation out on the numbers 3.01, 3, and −3.01 to see how it works.)

Another related function is that of the "nearest integer" to x, which we denote by <x>.[29] As the name implies, <x> is the integer

---

[29] The HP-67/97 has a RND key that returns <x> in the FIX, DSP 0 mode. Consult your owner's manual for details. Some TI calculators will round a number off to the number of digits in the display, using the sequence $\boxed{\text{EE}}$, $\boxed{\text{INV}}$, $\boxed{\text{EE}}$.

closest to x. Admittedly, this definition is ambiguous for numbers x that are exactly half way between two consecutive integers. For example, depending on the program you use, $<37.5> = 37$ or 38.

Problem 3: Write a program that accepts as input any number, x, and outputs $<x>$.

Solutions to Problems

Solution to Problem 1: Notice that if $\lfloor x \rfloor$ is not equal to INT(x), it is equal to INT(x) $-$ 1. Thus, $\lfloor 9.81 \rfloor = 9 =$ INT(9.81) and $\lceil -13 \rceil = -13$ $=$ INT(-13), whereas $\lfloor -9.81 \rfloor = -10 =$ INT(-9.81) -1. More precisely,

$$\lfloor x \rfloor = \begin{cases} \text{INT}(x) & \text{if INT}(x) \leq x \\ \text{INT}(x) - 1 & \text{if INT}(x) > x \end{cases}$$

(Recall that $\lfloor x \rfloor \leq x$ by definition.) The flowchart for the solution is shown in Fig. 1-35.

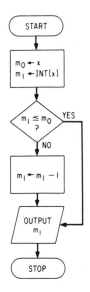

Fig. 1-35 Flowchart for inputting any number x and outputting $\lfloor x \rfloor$

Solution to Problem 2: As in Problem 1, $\lceil x \rceil$ often equals INT(x) but now sometimes equals INT(x) $+$ 1. More precisely,

$$\lceil x \rceil = \begin{cases} \text{INT}(x) & \text{if INT}(x) \geq x \\ \text{INT}(x) + 1 & \text{if INT}(x) < x \end{cases}$$

The flowchart for the solution is shown in Fig. 1-36.

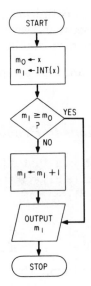

Fig. 1-36 Flowchart for inputting any number x and outputting [x]

Solution to Problem 3: Either one of two possible formulas will work:

$$\langle x \rangle = \lfloor x + .5 \rfloor$$
$$\langle x \rangle = \lceil x - .5 \rceil$$

Using the first, for example, $\langle 37.5 \rangle = 38$; using the second, $\langle 37.5 \rangle = 37$. For numbers x *not* exactly half way between consecutive integers, the two formulas give the same answer. To produce a flowchart, either take the flowchart from Problem 1 and replace the step, $m_0 \leftarrow x$, by the step, $m_0 \leftarrow x + .5$, or take the flowchart from Problem 2 and replace the step, $m_0 \leftarrow x$, by the step, $m_0 \leftarrow x - .5$.

Notes: The function "fractional part" of x appears on most calculators. The notation on the HP is FRAC; on the TI, INV INT. It is defined by the equation, $FRAC(x) = x - INT(x)$. Thus, $FRAC(3.05) = .05$, $FRAC(3) = 0$, and $FRAC(-3.05) = -.05$.

1.13 Changing Bases

Difficulty: 3

The numbers we ordinarily calculate with are written in the decimal, or base 10, number system. As you know, this system represents any positive whole number by a digit or a sequence of digits represented by the symbols 0, 1, 2, 3, 4, 5, 6, 7, 8, and 9. The digit at the right end of the sequence counts the number of 1s, the next digit to the left counts the number of 10s, the next digit to the left counts the number 100s, and so forth. Since $1 = 10^0$, $10 = 10^1$, $100 = 10^2$, etc., we can say that the symbol in the ith position (reading from right to left, and calling the rightmost position the 0th one) counts the number of 10^is. Thus, $356 = (3) (10^2) + (7) (10^1) + (6) (10^0)$. In general, $a_k a_{k-1} \ldots a_1 a_0$ represents

(1) $(a_k) (10^k) + (a_{k-1}) (10^{k-1}) + \ldots + (a_1) (10^1) + (a_0) (10^0)$

Let it be noted that there is nothing special about the base number 10; any whole number b, where $b \geq 2$, can be used as a base. We can illustrate this fact by considering base 8 ($b = 8$), the so-called octal system. Each number in the octal system is represented by a digit or a sequence of digits represented by 0, 1, 2, 3, 4, 5, 6, and 7. (Notice, as in the decimal system, that we stop one number short of the base.) The rightmost digit represents the number of 1s ($1 = 8^0$), the next the number of 8s ($8 = 8^1$), the next the number of 64s ($64 = 8^2$), and so on. For example, $653_8 = 427_{10}$ since $653_8 = (6) (8^2) + (5) (8^1) + (3) (8^0) = 427$. (The subscript 8 in 653_8 indicates that this is an octal number just as the subscript 10 in 427_{10} indicates a decimal number. In general, then,

(2) $(a_k a_{k-1} \ldots a_1 a_0)_8 = a_k 8^k + a_{k-1} 8^{k-1} + \ldots + a_1 8^1 + a_0 8^0$

Problem 1:   Write a program that accepts as input a base 8 number and outputs its base 10 value.

If s is any integer in the range, $2 \leq s \leq 10$, your solution to Problem 1 can be readily adjusted to convert from base s to base 10 by replacing all occurrences of the number 8 by the number s.

Now, how do we convert from base 10 to base s? The answer is remarkable. Merely take your solution to Problem 1, replace all occurrences of 10 by s, and replace all occurrences of s by 10. The resulting program will convert from base 10 to base s!

Suppose that s and t are integers between 2 and 10, inclusive. How do we write a program to convert from base s to base t? In view of the comments above, you might think that using the solution to problem 1, with 8 replaced by s and 10 replaced by t, will do the job. Unfortunately, this is *not* the case, unless either s or t happens to be equal to 10. You will have to be a bit more clever.

Problem 2:   Write a program that converts numbers from base s to base t, where s and t are any two integers between 2 and 10. The program will be initialized by storing s and t in a pair of memories.

Solutions to Problems

Solution to Problem 1:   We will use three memories—m_0, m_1, and m_2. Memory m_0 initially stores the input number in base 8. Digits are fished out of the right end of m_0 one by one so that, for example, if the input is 653, m_0 first contains a 653, then a 65, then a 6, and finally a zero, as the digits 3, 5, and 6 are removed. Memory m_1 initially contains a 1 and then the successive powers, 8^1, 8^2, 8^3, etc., needed to evaluate Formula (2). Memory m_2 starts at zero and accumulates the sum in Formula (2), working from right to left. In the example above $(653_8 = 427_{10}$, m_2 first contains a zero; then $3 = 3 \cdot 8^0$; then, $43 = 5 \cdot 8^1 + 3 \cdot 8^0$; and finally the answer, $427 = 6 \cdot 8^2 + 5 \cdot 8^1 + 3 \cdot 8^0$.

The flowchart for this solution in shown in Fig. 1-37. The first

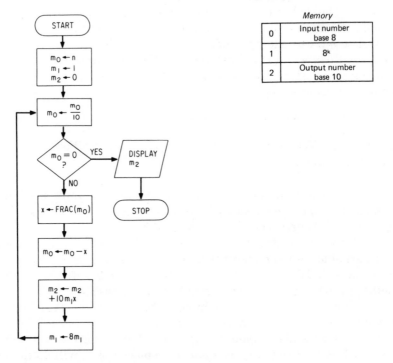

Memory	
0	Input number base 8
1	8^k
2	Output number base 10

Fig. 1-37   Flowchart for inputting a number in base 8 and outputting it in base 10

box initializes everything. The next box divides m_0 by 10, thereby sliding the string of digits in m_0 to the right by one decimal place and pushing the rightmost digit one place to the right of the decimal point. The next box tests to see if $m_0 = 0$ (that is, tests to see if we have used all the digits in m_0) and displays the answer if we have. If $m_0 \neq 0$, we are not done yet and must continue. The step, $x = FRAC(m_0)$, recovers the one digit to the right of the decimal point, while the step, $m_0 \leftarrow m_0 - x$, removes this last digit from m_0. Since the digit in x is one digit to the right of the decimal point, it must be multiplied by 10 to produce the actual digit. The step, $m_2 \leftarrow m_2 + 10m_1x$, then adds to m_2 the digit $10x$, multiplied by the appropriate power of 8 in m_1 in accordance with Formula (2). The step, $m_1 \leftarrow 8m_1$, increases the power of 8, and we then go back to the top and do the next digit.

Solution to Problem 2: The basic idea is that if a and b are integers between 2 and 10, inclusive, then the flowchart for Problem 1, with 8 replaced by a and 10 by b, will convert from base a to base b *only* if either a or b equals 10. Thus, to convert from base s to base t, we first convert from base s to base 10, using the proceeding routine where $a = s$ and $b = 10$; then we convert the resulting output in base 10 to base t, again using this routine where $a = 10$ and $b = t$.

Since the same routine is used twice, we use a subroutine for it (see Sec. 1.7). The subroutine requires two more registers, m_3 and m_4, to store a and b, respectively. The main program is initialized by storing s in m_5 and t in m_6.

The flowchart for converting n from base s to base t is shown in Fig. 1-38.

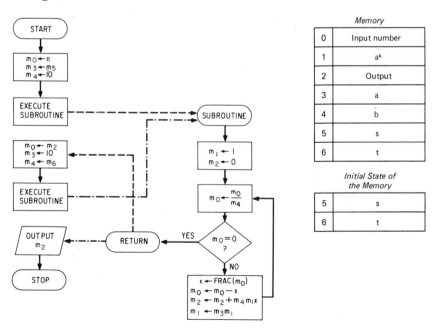

Fig. 1-38 Flowchart for inputting a number in base s and outputting it in base t

1.14   Multiple Storage

Difficulty: 2–3

Your calculator probably has between 8 and 30 addressable memories (unless you happen to have a TI 58/59 or HP-41C). For many programs, even eight memories are enough, but what if you need more? In this section, we will show you how to get the most of your memories, whatever their number. The basic trick to coerce each of your memories into storing several numbers at once. For example, if your memories have 10 significant digits each, two five-digit numbers can be stored in one memory. The easiest way to do so is to store one number in the five digits to the left of the decimal point and the other in the five digits to the right.

Problem 1:   Write programs to store and recall two nonnegative integers, each with at most five digits, using only one memory as described above.

We can take the idea even further. Assuming that each memory can hold ten digits, it can be divided up in many ways besides two chunks of five. If the only numbers to be dealt with are the digits 0, 1, 2, . . ., 9, say, then 10 such numbers will fit into a single memory. If the numbers range from 0 to 99, five of them can be stored in one memory, each occupying a two-digit chunk of the ten-digit memory.

Problem 2:   Write programs to store and recall, using one memory, the following:

   (a) Ten integers, each between 0 and 9
   (b) Five integers, each between 0 and 99
   (c) Three integers, each between 0 and 999

Here is one final technique. Suppose that we wish to store a bunch of numbers, all either 0 or 1. Using Problem 2(a), we can certainly get 10 such numbers per memory, but there is a way to pack in at least 30 that uses binary (base 2) numbers (if you haven't studied Sec. 1.13, you should do so now before reading on).

Consider, for example, these 30 zeros and ones in a row:

$$000100101010001101100100000111$$

Doesn't this look suspiciously like a base 2 number with 30 digits? The largest such a number can be is $2^{30} - 1$ or 1073741823, which occurs when all digits are a 1 in base 2. This string of zeros and ones, then, can be stored in a single memory as a base 10 number no larger than 1073741823.

Problem 3:   Write programs to store and recall 30 zeros or ones using a single memory.

Other bases, of course, can be used.

Problem 4:   Write a program to store and recall, using one memory, the following:

(a) Seventeen numbers, each either a 0, 1, or 2
(b) Fifteen numbers, each either 0, 1, 2, or 3
(c) Thirteen numbers, each either 0, 1, 2, 3, or 4
(d) Eleven numbers, each between 0 and 6

Use base 3 for (a), base 4 for (b), base 5 for (c), and base 7 for (d). (We will not give a solution for this problem; you are on your own.)

Solutions to Problems

Solution to Problem 1:   We will use a single register, m_0, to store our numbers. Now think of m_0 as split into two halves, ℓ and r (for left and right), by the decimal point, as follows:

$$\underset{\text{DECIMAL}}{\underline{\quad\quad\ell\quad\quad} \;.\; \underline{\quad\quad r\quad\quad}}$$

To store in ℓ a nonnegative whole number x with at most five digits, without disturbing r of course, use

$$m_0 \leftarrow x + \text{FRAC}(m_0)$$

The operation $\text{FRAC}(m_0)$ "clears" the ℓ register and, by adding x to the result, places x in the cleared ℓ register. To recall ℓ is even easier; just calculate $\text{INT}(m_0)$.

Storing and recalling r are a little more complicated. The flow-charts for these operations are shown in Fig. 1-39. The multiplications

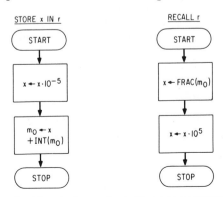

Fig. 1-39 Flowchart for storing and recalling two nonnegative integers

by 10^{-5} and 10^5 shift the decimal point five places. (*Caution:* you might want to use your 10^x or INV log x key to calculate 10^5 and 10^{-5} but check to see whether they give *exact* answers before using them (see Sec. 1.10). Another possibility is to have the precise values of these numbers—100,000 and .00001—stored in memory as in Sec. 1.2(7).

Solution to Problem 2:   Since the three parts are basically the same, we will describe problem (c) only and leave the others to you. Let us divide memory m_0 into three three-digit chunks, as follows:

Here are the formulas for storing x:

Store a: $m_0 \leftarrow 10^6 [x + \mathrm{FRAC}(10^{-6}\ m_0)]$
Store b: $m_0 \leftarrow 10^{-3}\ (m_0)$
              $x \leftarrow x + \mathrm{FRAC}(m_0)$ (the x to the right of the $\leftarrow$ is the x to be stored)
              $m_0 \leftarrow 10^3 [10^3\ \mathrm{INT}(10^{-3}\ m_0) + x]$
Store c: $m_0 \leftarrow 10^3\ \mathrm{INT}(10^{-3}\ m_0) + x$

The formulas for recalling x are as follows:

Recall a: $x \leftarrow \mathrm{INT}(10^{-6}\ m_0)$
Recall b: $x \leftarrow \mathrm{INT}\ [10^3\ \mathrm{FRAC}(10^{-6}\ m_0)]$
Recall c: $x \leftarrow \mathrm{INT}\ [10^3\ \mathrm{FRAC}(10^{-3}\ m_0)]$

Why these six formulas work will become clear if you go through one example by hand.

In this solution, the numbers 10^{-6}, 10^{-3}, 10^3, and 10^6 are needed. You could use your 10^x (INV log x on TI calculators) to calculate them, but that key might not give the exact answer. If it doesn't, you will have to key the constants in directly. Another approach is to trick the 10^x into being accurate (see Sec. 3.17).

Suppose that we wish to store nine digits, $a_1, a_2, \ldots, a_9$, each between 0 and 9, in a single register, m_0. There are many ways to do so, for example,

$$\underline{|\ a_9\ |\ a_8\ |}\ \ldots\ \underline{|\ a_2\ |\ a_1\ |\ 0\ |}.$$

Suppose further that we want to write a program that will first accept as input an integer i between 1 and 9 and then increase a_i by 1 (without messing with the other a's, of course). The following step ought to do it:

$$m_0 \leftarrow m_0 + 10^i$$

Since we assume that a_1 is not already 9, there is no carryover in adding $m_0 + 10^i$.) If your 10^x key returns *exactly* 10^i for $1 \le i \le 9$, the program works fine. But what if it doesn't?

In such a case, we change our strategy, storing the a's to the right of the decimal point and placing an extra digit 1 in the unit's place:

$$m_0 = 1.a_1 a_2 a_3 a_4 a_5 a_6 a_7 a_8 a_9$$

Now we use the revised formula

$$m_0 \leftarrow m_0 + 10^{-i}$$

We may ask why this formula works whereas the previous one did not. Let us try an example when $m_0 = 1.000000000$ and $i = 8$.

On the HP-25, we find that using 10^x where $x = -8$ gives: $10^{-8} = 1.000000004 \times 10^{-8}$. Thus, $m_0 + 10^{-8} = 1.00000001000000004$, which the calculator rounds off to 1.000000010, the correct answer.

Notice what the artificial 1 to the left of the decimal does. Were it not there, the calculator would store .00000001000000004 as $1.000000004 \times 10^{-8}$ and fail to round off the 4 at the end.

Solution to Problem 3: If the 30 numbers are $a_0, a_1, a_2, \ldots, a_{29}$, let us store them as follows: $m_0 = a_{29}a_{28} \ldots a_2 a_1 a_0$ (base 2). Here is the formula for recalling a_i, for any i between 0 and 29, inclusive:

$$x \leftarrow INT[2\ FRAC(m_0/2^{i+1})]$$

To store x in the ith position is a bit harder. Here's how:

$$m_0 \leftarrow (m_0/2^i)$$
$$x \leftarrow x + FRAC(m_0)$$
$$m_0 \leftarrow 2^i[2\ INT(1/2\ m_0) + x]$$

The best way to understand these formulas is to pretend that all numbers are base 2; then multiplying or dividing by 2^i just shifts the "decimal point." (We should really call it the binary point.)

1.15 Searching

Difficulty: 3

In 1940, world population was 2.249×10^9. In 1950, it was 2.509×10^9. If the growth rate represented by these figures remains constant, then, t years after 1940, world population will be given by the expression, $P = 2.249 \times 10^9 (1.116)^{t/10}$. Suppose that you want to use this formula to find the year in which world population will reach five billion (5×10^9). This is an example of what is called an *inverse problem*. The formula is set up to go one way. Given a time t, it readily returns the value of P at that time. For example, when t = 10 (in 1950), P = 2.249 × 10⁹ (1.116)¹ = 2.249 × 10⁹ × 1.116 = 2.51 × 10⁹.[30] The question, however, calls for us to go the other way. Given a value of P (5 billion), we are asked to find t.

One way to solve the problem is to invert the equation, that is, solve it for t in terms of P. In this case, the equation can be inverted without too much difficulty (if you know the trick; see the Notes). But if you are unable to solve such an equation, you will have to *search* for a solution.

To solve the problem at hand, program your calculator to return values of P for given values of t [test your program first by trying it for t = 0 (1940) and t = 10 (1950)]. By experimenting around, you will discover that the desired value of t lies somewhere between 72 and 73 (that is, somewhere in the year 2012). This answer is sufficient for the present problem. Under other circumstances, however, you might want to determine t accurate to five decimal places. You could do so by zeroing in on the answer from above and below, but that method would be tedious.

What we want to describe in this section is a procedure for zeroing in on a solution that can be programmed onto your calculator in a few steps but is nevertheless very efficient. It is called a *binary search*.

To set the stage, let us say you have a formula, $Q = f(x)$. The righthand side of the equation is some mathematical expression that you can program on your calculator. Given a value of Q, you are asked to find that value of x which yields this particular Q. (As a specific example, letting $Q = x \ln x$, find the value of x for which Q = 5280.) We want to produce a program into which the program for computing values of Q can be embedded and that will search out the value of x required. It will zero in on x from above and below.

Let's call the given value of Q the *target value*. In order to start a binary search, you need a value of x, say x_0, for which the corresponding value of Q is *smaller* than the target value, and another value of x, say x_1, for which it is *larger*. The value of x we are looking for is then somewhere between x_0 and x_1.[31] We will make a stab at the desired x by taking it to be exactly half way between x_0 and x_1. Expressed mathematically, $x_m = (x_0 + x_1)/2$. If Q is too large at x_m, then let x_m be the new x_1. If Q is too small at x_m, then let x_m be the new x_0. In this way, the search area for the proper value of x is cut in half

---

[30] You can see that rounding off (2.509 × 10⁹)/(2.249 × 10⁹) to three decimal places (1.116) introduce errors in the formula; we should have obtained 2.509 × 10⁹

[31] At least it will be with any decent luck (see the Notes).

(since x is now trapped between x_0 and x_1, one of these having been changed to the x_m defined above, the distance between x_0 and x_1 is half what it was formerly). By repeating this procedure, we can zero in on x as closely as we like.

Problem: Write a program that carries out the above procedure. The program should provide space for the user to include the rule for computing values of Q (x_0, x_1, and the target value of Q will be stored in memories). When $\boxed{\text{R/S}}$ (or whatever) is pushed, the program should stop with x_m in the display. When $\boxed{\text{R/S}}$ is pushed again, it should stop with the value of Q at x_m in the display (having meanwhile moved x_m into the appropriate memory). In this way, the user can continue pressing $\boxed{\text{R/S}}$ until a value of Q sufficiently close to the target value is reached or until x_m varies within acceptable bounds.

Once you have solved this problem, you may want to get fancy and automate the search further. Set things up so that the user can specify how accurately x and the target value of Q are to be approximated. The program will then recycle itself until the desired level of accuracy is achieved.

Solution: Four memories will be used: m_0 will hold the number x_0, m_1 will hold x_1, m_2 will be temporary storage for x_m, and m_3 will hold the target value of Q. Since the program is essentially sketched in the statement of the problem, we can proceed directly to the details. The flowchart is shown in Fig. 1-40.

The program begins by computing x_m and stops with this number in the display, thereby showing the user the value of x for which Q is about to be computed. Before (or immediately after) the $\boxed{\text{R/S}}$ order, x should be stored in m_2. Next the program computes Q from x. This part of the program is to be supplied by the user, and if your calculator permits subroutines, you will want to make it subroutine. If not, just leave space in the big program for the user to key in the program for computing values of Q. In either case, the user should test his program for Q and determine the initial values of x_0 and x_1 to be stored in m_0 and m_1 before running the big program.

After Q has been computed from x_m, the program compares Q with the contents of m_3. If Q is greater than the target value, x_m (which is in m_2) is stored in m_1. This will be the new value of x_1. Note that the difference between x_0 and x_1 has been cut in half. If Q is not greater than the target value, the program then stores x_m in m_0, and this becomes the new value of x_0. (Be careful not to lose Q during this testing procedure, for you want to end the program with Q in the display; you may have to store it temporarily.) With Q in the display, the program should end with R/S followed by RST (TI) or end with

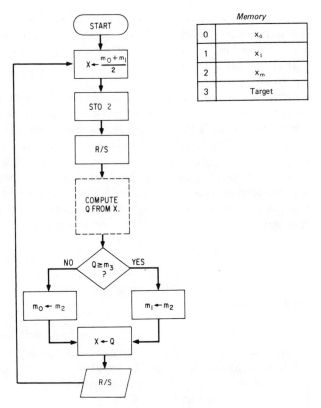

Fig. 1-40 Flowchart for binary search

GTO 00 or RTN or INV SBR (HP) so that the next time the user pushes $\boxed{\text{R/S}}$ the program will return to the top and compute another value of x_m. If your calculator permits subroutines, you may prefer simply to make the whole program a subroutine to be called each time the user wants a new x_m and Q. [*Solution was realized on an HP-25 in 17 steps.*]

The program can be further automated as indicated in the statement of the problem. Suppose, for example, that you want to find a value of x yielding the target value of Q with four-digit accuracy. The way the program is presently set up, the user will have to press $\boxed{\text{R/S}}$ until this level of accuracy is obtained. The program can be modified, to do this on its own, as follows: First, store 10^{-5} in m_4. Then, at the end of the program, instead of stopping, compute the absolute value of the difference between Q and the target value and compare this with m_4. If m_4 is smaller, you haven't obtained the desired level of accuracy and must return control to the top of the program for the next value of x_m. You also will want to remove the R/S from the middle of the program so that the program will keep going until m_4 finally exceeds the difference between Q and the target value. When this happens, stop the program with Q in the display.

A similar test can be added at the end if you want to get x_m within a certain number of digits. This time you will test the absolute value of the difference between x_0 and x_1 against the contents of some memory specifying the desired level of accuracy. The two tests can be linked in series if you want to control both x and Q.

These tests should not be added to your program (except perhaps for the sake of practice) if you have a small calculator. The reason is that you want to save all the space you can for the user to program in the formula for Q.

Notes: Although unlikely, it could happen that no value of x exists between x_0 and x_1 that yields a Q equal to the target value. This situation may occur if Q is undefined somewhere between x_0 and x_1. For example, suppose that Q = 1/x. Even though Q = −1 at −1 and Q = 1 at 1, there is no x between −1 and 1 (or anywhere else) for which Q = 0. (Observe that Q is undefined at x = 0).

If you want to automate still more of your search procedure, you could construct a program to search for initial values of x_0 and x_1, thus saving the user the trouble of doing so, but it is probably more trouble than it is worth. Since every search must start somewhere, why automate a search for the starting point of a search? Where will this search start?

Our original example can be solved without a calculator search. We want to find value of t for which P = 5 × 10⁹, where P is given by the formula, $P = 2.249 \times 10^9 (1.116)^{t/10}$. Hence, we want to solve the equation, $5 \times 10^9 = 2.249 \times 10^9 (1.116)^{t/10}$, for t. There is a general rule for solving such equations, in which the unknown appears as an exponent: Take logs on both sides of the equation.[32] In the present instance, this yields the equation, $\log(5 \times 10^9) = \log(2.249 \times 10^9 (1.116)^{t/10})$, which, by applying the rules of logarithms [see Sec. 1.2(8)] simplifies to the equation, $\log(5) + 9 = \log(2.249) + 9 + (t/10) \log(1.116)$. The latter equation can now be fairly easily solved, as follows: $t = 10(\log 5 - \log 2.249)/\log(1.116)$, or 72.8 (use your calculator).

1.16   Space and Time in the Calculator

Difficulty: 4

In this section, we shall describe how to measure the efficiency of a program. Since many programs will be capable of solving any given problem in this book, you should be capable of evaluating your own—with the aim, of course, of improving it.

---

[32] The converse of this rule says that to solve an equation involving the log of the unknown, take antilogs (10^x or e^x) on both sides.

Two important aspects of a program are *space* and *time*. Space, which refers to the amount of storage used by the program, is of two kinds: *program space* (that is, the number of program steps used) and *register space* (that is, the number of storage registers needed). Space is usually much more critical on a pocket calculator than on a computer. The largest pocket calculators today provide for about a thousand program steps and close to a hundred registers, but even minicomputers may offer a hundred times more space.

Time, which refers to the time required to run the program, is not usually measured in seconds or hours, but rather in the number of basic operations the program must execute. For example, consider the following problem: For every positive integer n, calculate

$$s_n = 1 + 2 + 3 + 4 + \ldots + (n - 1) + n$$

A sample program to solve this problem is shown in Fig. 1-41. Register m_1 successively contains $0, 1, 2, \ldots , n$; register m_2 stores the partial sums: $s_0 = 0$; $s_1 = 1$; $s_2 = 1 + 2$; $s_3 = 1 + 2 + 3 \ldots , s_n = 1 + 2 + 3 + \ldots + (n - 1) + n$.

The program requires approximately 20 steps and uses only three registers, m_0, m_1, and m_2. Hence, the *space* it requires is quite small. To measure the *time* it requires, we must first determine its basic operations. By examining the flowchart, we see that there are four of them: storing and recalling numbers, incrementing (adding 1 to) a number, adding two numbers together, and comparing two num-

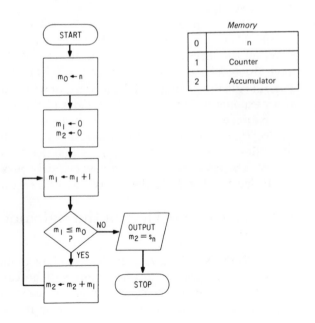

Fig. 1-41 Flowchart for calculating s_n where $s_n = 1 + 2 + 3 + \ldots + (n - 1) + n$

bers ($m_1 \leq$ n?). Let us ignore (for the sake of simplicity) the first operations and concentrate on incrementing, adding, and comparing.

Since register m_1 begins at 0 and ends at (n + 1), it is incremented (n + 1) times. Register m_2 has n additions performed on it, and the comparison, $m_1 \leq$ n?, is used (n + 1) times by the program. Altogether, we use each of our three basic operations about n times. Evaluating s_{10000}, therefore, requires about 30,000 steps—100 times the time required to evalute s_{100} (300 steps)!

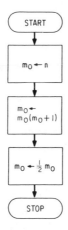

Fig. 1-42 Flowchart for calculating s_n by calculating ½n(n + 1)

Consider another program for solving the same problem (Fig. 1-42). This program calculates 1/2 n(n + 1). What has this quantity to do with our problem? Well, in fact,

$$1 + 2 + 3 + \ldots + n = 1/2\ n(n + 1)$$

For a short proof, note that $s_n = 1 + 2 + \ldots + (n - 1) + n$ and $s_n = n + (n - 1) + \ldots + 2 + 1$. Adding the two equations, we get

$$2s_n = (n + 1) + (n + 1) + \ldots + (n + 1) + (n + 1)$$

Since there are n terms added on the right-hand side, each term equal to n + 1, $2s_n = n(n + 1)$, or

$$s_m = 1/2\ m(n + 1)$$

In analyzing this program, we find that it uses fewer steps (at most 10) and fewer memories (only 1) than Program 1. It therefore saves more space, but, more important, it takes far less time. The program increments only once (to calculate n + 1) and performs only one multiplication and one division (by 2). Only three basic operations are performed, no matter how large n is. Thus, the calculation of s_{10000}

takes virtually the same length of time as the calculation of s_{100}.[33] This great contrast in the time required by the two programs shows that the second is definitely more efficient than the first.

Let us examine another problem. For each whole number x in the range, $0 \leq x \leq 99$, let f(x) denote the last two digits of the product of x multiplied by 14 (that is, 14x) so that f(x) also conforms to the range, $0 \leq f(x) \leq 99$. For example, f(0) = 0, whereas f(17) = 38 (since $14 \times 17 = 138$ and the last two digits are 38). Now, starting with the number 3, apply this operation repeatedly to generate the following list in which every number is the f(x) of the previous number x:

3, 42, 88, 32, 48, 72, 8, 12, 68, 52, 28, 92, 88, 32, 48, . . .

Notice that the block of ten numbers—88, 32, . . . , 28, 92—repeats over and over again, whereas the first two entries, 3 and 42, never reappear.

We can visualize this list as shown in Fig. 1-43. Here twelve points are labeled with the twelve different numbers occurring in the list, and an arrow is drawn from each point to its f(x) value. As this diagram resembles the greek letter ρ (rho), we call it the ρ-*diagram* of f(x) starting at 3. Let us denote by c the number of points on the circular part of the ρ-diagram, and by t the number of points in the tail of the ρ-diagram. In Fig. 1-43, c = 10 and t = 2. The point where the tail enters the circle is called the *leader* of the ρ-diagram; thus, 88 is the leader of the present ρ-diagram.

This type of set-up occurs often (see Secs. 2-11 and 3-9, for example). The general statement of the problem is as follows: Given a finite collection, F, of numbers, and a function, f, assigning to each x in F a number, f(x), also in F; and, given a number, $x_0 \in F$, (1) find the numbers, c and t, associated with the ρ-diagram of f starting at x_0, and (2) find the leader. (In the preceding example, F represents the collection of the 100 whole numbers x in the range $0 \leq x \leq 99$; f(x) represents the last two digits of the quantity, 14x; and $x_0 = 3$.)

We will discuss and analyze three programs for solving this problem. The first, the naive method, will be impractical in most cases. The second is not really a method at all but offers some good ideas, some of which are used in the third, the Pollard ρ method. This approach was first used by its originator in a classical number theory problem to determine whether a number is prime. It is an excellent example of how far a bit of cleverness can go in improving the efficiency of a program.

---

[33] The second program uses three basic operations to calculate either s_{100} or s_{10000}. However, your calculator will take slightly more time to calculate x_{10000} than s_{100}. The reason is that the numbers involved in calculating s_{10000} are much bigger than those in calculating s_{100}.

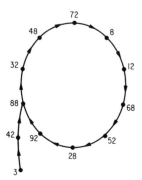

Fig. 1-43 The ρ-diagram

Naive Method

For the series, $x_1 = f(x_0)$, $x_2 = f(x_1)$, $x_3 = f(x_2)$, etc., we define x_1 as the result of applying f to x_0 i times. The naive method consists of the following: We calculate x_0, x_1, x_2, etc., one after the other and store them in memories m_0, m_1, m_2, etc., as they are computed. As each new x_1 is computed, we test whether $x_1 = x_{1-1}$?, $x_1 = x_{i-2}$?, and so forth, until we reach the test, $x_i = x_0$? If none of the tests pass (that is, if x_i is a number not equal to any of the earlier x's), then x_{i+1} is computed, stored in m_{i+1}, and the process repeated. Otherwise, if $x_i = x_j$ for some j in the range, $0 \leqslant j \leqslant i - 1$, we are done. We have found the first repeating number in the diagram and consequently $t = j$, $c = i - j$, and x_j is the leader. (In the example above, $x_2 = x_{12} = 88$ is the first repetition; consequently $t = 2$, $c = 12 - 2 = 10$, and 88 is the leader).

A flowchart for the naive method is shown in Fig. 1-44. This flowchart actually does execute the naive method. If you cannot see why, try working through it with a pencil and paper using the specific example where $f(x) =$ the last two digits of $14x$ and $x_0 = 3$. Now to the analysis!

The amount of program space used here is quite reasonable. If your calculator has indirect addressing, no more than 30 or 40 program steps are required. The amount of storage space needed, on the other hand, is outrageous. Besides the two registers, i and j, we need separate registers, m_0, m_1, . . . , m_{c+t}, for each of the numbers, x_0, x_1, . . . , x_{c+t}. This is a total of $c + t + 3$ registers, or $k + 3$ registers, where k is the total number of points, $c + t$, in the ρ-diagram of f starting at x_0.

This is the main reason the naive method is impractical. Even the simple example with $c = 10$ and $t = 2$ requires 15 registers (10 + 2 + 3). It will obviously not work on small calculators with only 8 registers; and it is easy to find examples that will not work on much

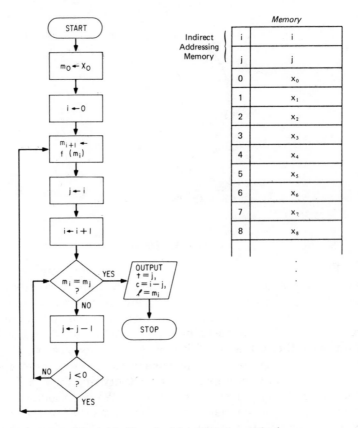

Fig. 1-44 Flowchart for the naive method

larger calculators. The ρ-diagrams produced by random-number generators (see Sec. 3-9) can easily have a c greater than 1000!

To analyze the time taken by the naive method, take note of the basic operations involved: storing, recalling, incrementing ($i \leftarrow i + 1$), decrementing ($j \leftarrow j - 1$), comparing ($m_i = m_j?; j < 0?$), and using the function f [$m_{i+1} \leftarrow f(m_i)$]. For the sake of simplicity, however, we shall deal with only two of these: the comparison, $m_i = m_j$, and using the function f.

Since each of the numbers, $x_1 = f(x_0)$, $x_2 = f(x_1)$, ..., $x_k = f(x_{k-1})$, must be computed, the function f is used exactly k times. As for the comparisons,

1. x_1 must be compared with x_0
2. x_2 must be compared with x_1 and x_0, and so on, until
3. x_{k-1} must be compared with x_{k-2}, x_{k-3}, ..., x_1, x_0, and
4. x_k must be compared with x_{k-1}, x_{k-2}, ..., x_t

The total number of comparisons is thus $[(1 + 2 + 3 + \ldots + (k - 1)]$ $+ (k - t)$, or $(1 + 2 + 3 + \ldots + k) - t$. By the same reasoning used previously, $1 + 2 + 3 + \ldots + k = 1/2 \, k(k + 1)$. The total number of comparisons is therefore $1/2 \, k \, (k + 1) - t$.

Here is a summary of our analysis of the naive method:

Number of storage registers used $= k + 3$
Number of uses of the function $f = k$
Number of comparisons $= 1/2 \, k(k + 1) - t$

Pseudo-Method

The second method uses only a few storage registers, but it has drawbacks, as you will see. First, let's program an infinite loop to calculate the points, x_0, x_1, $= f(x_0)$, $x_2 = f(x_1) = f[f(x_0)]$, $\ldots$, etc., in the ρ-diagram of f at x_0. We start the program off and wait a long time for i to get so large that x_i is on the cycle (and off the tail) of the ρ-diagram. When we think we have waited long enough, we stop the program and examine the current x_d. If it is on the cycle, then $x_d = x_{d+c}$, and this information can be used to find c with a second program. This program permanently stores x_d in some memory and sets up a loop that successively calculates the numbers, $x_{d+1}, x_{d+2}, x_{d+3}, \ldots, x_{d+j}$, etc. Each time through the loop, the current x_{d+j} is tested to see if $x_d = x_{d+j}$. If not, the program goes back to the top of the loop, but if $x_d = x_{d+j}$, then $c = j$ and we have found c. (Our loop must have a counter to store the current value of j at the jth time through the loop (so that the counter is incremented by 1 every time through.)

Now that we have found c, it is easy to find t and the leader x_t. We use the fact that t is equal to the smallest i for which $x_i = x_{i+c}$. First we calculate x_c by setting up a loop to calculate successively x_1, x_2, x_3, $\ldots$, etc., and stopping after exactly c iterations. There is an easier way, though. The loop previously used for finding c is iterated exactly c times; thus we can make it do some extra work. Let an additional register initially store x_0 and apply the function f to it each time through the loop. After c iterations, it will contain x_c.

Having found x_c, we set up a loop with two memories, initialized by x_0 and x_c, to find t. Each time through the loop, the function f is applied to both registers; at the ith time through, one register will contain x_i and the other x_{i+c}. At the end of the loop, we test to see if $x_i = x_{i+c}$. If it does not, we go back to the top of the loop. If it does, then $i = t$ and the point, $x_i = x_{i+c}$, is the leader. (For flowcharts of all these routines, see solution to the problem for the Pollard ρ-method.)

The drawback to this second method is that it really isn't a method at all. The crucial step occurs when we "wait a long time."

How long should we wait? If too long, we are wasting time. If not long enough, we are still in the tail of the ρ, and the second phase of the program will never' stop because the test, $x_d = x_{d+i}$, will never be passed.

Pollard ρ-Method

Finally, we come to the Pollard ρ-method. All the troubles in the second method were concentrated on finding any x_d at all *in the circular part* of the ρ. Here is the stroke of genius that finds such an $\dot{x}_d$. In the flowchart of Fig. 1-45, only two registers, m_0 and m_1, are used, each containing x_0 at first. Each time through the loop, m_0 is replaced by $f[f(m_0)]$, and m_1 is replaced by $f(m_1)$. After i times through the loop, then, m_0 contains x_{2i} and m_1 contains x_i. If the program stops after d runs through the loop, then $x_d = x_{2d}$ and must therefore be an element in the circular part of the ρ. The reason for this is that no element in the tail of the ρ equals any other point in the entire ρ-diagram; consequently, the point, $x_d = x_{2d}$, cannot be on the tail. Thus, if the program does stop, it will have found a required x_d.

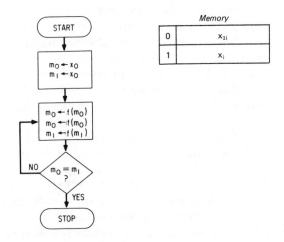

Fig. 1-45 Flowchart for Pollard ρ-method

Can the program go on forever? It cannot; in fact, the number of runs through the loop is at most $k(= c + t)$. Let's take a look at the ρ-diagram again. The program generates two sequences of numbers, $x_i = x_1, x_2, x_3, \ldots$, and $x_{2i} = x_2, x_4, x_6, \ldots$. The two sequences move along the the ρ-diagram starting from x_1 and x_2 as shown in Fig. 1-46.

Moving twice as fast as x_i, x_{2i} stays in front. After t iterations, x_i enters the circular part of the ρ, and x_{2i} is already there. Now x_i and x_{2i} move around the circle, x_i one step at a time and x_{2i} two. With x_{2i}

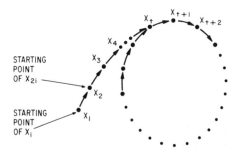

Fig. 1-46 Movement of the two sequences, x_i and x_{2i}

gaining one step each time on x_i, it eventually catches up with x_i; at this point, when $x_{2i} = x_i$, the program stops. How long does it take? It takes t steps for both x_i and x_{2i} to get into the circle. After that, x_{2i} has to advance no more than c steps on x_i to catch it. (Since there are c steps in the entire circle, you will see that x_{2i} actually need advance no more than c − 1 steps.) Thus, $x_i = x_{2i}$ after no more than c + t steps. Once x_d is located, it is relatively easy to find c, t, and the leader ℓ, as we did above.

Problem: Write a program using the Pollard ρ-method that accepts as input the number x_0. It should return the numbers c, t, and ℓ in the ρ-diagram of f starting at x_0.

Solution: The solution has three phases:

1. Find any d for which x_d is in the circular part of the ρ-diagram.
2. Find c and x_c.
3. Find t and ℓ.

Phase 1 was described just prior to the statement of Problem 1. Phases 2 and 3 were described in the discussion of the second method. Here are the details.

We use five registers, m_0 through m_4. To initialize, we place x_0 in m_1 through m_4, and 0 into m_0, which is our counter for the second and third phase. Each time through the loop of phase 1, only m_1 and m_2 are affected; f is applied once to m_1 and twice to m_2. The loop is repeated until $m_1 = m_2$ and we have found a point, $x_d = m_1 = m_2 = x_{2d}$, in the circular part. Each time through the loop of phase 2, counter m_1 is incremented, f is applied to both m_1 and m_4, and we test to see if $m_1 = m_2$, repeating the loop of phase 2 until $m_1 = m_2$. Since m_1 begins at x_d and m_2 remains at $x_{2d} = x_d$ throughout phase 2, the loop will take c iterations, ending with $m_0 = c$, $m_1 = x_{d+c} = x_d = m_2$, $m_3 = x_0$, and m_4 (which began phase 2 at x_0) now at x_c.

Before we begin phase 3, we place c (in m_0) in m_1 and x_0 (in m_3) also in m_2. We then reinitialize counter m_0 to 0. Each time through the loop of phase 3, we increment counter m_0, apply f to m_3 and m_4, and test to see if $m_3 = m_4$, continuing the loop until $m_3 = m_4$. Hence, m_3 will go from x_0 to x_t, and m_4 from x_c to x_{c+t}.

Table 1-1 summarizes how the five registers are used in the three phases. The arrows, $\xrightarrow{f}$, $\xrightarrow{ff}$, $\xrightarrow{\text{Increment}}$, mean that at each time through the loop, the contents x of the given register are replaced by f(x), f[f(x)], and x + 1, respectively, and a dashed arrow, $--\rightarrow$, means that the number is unchanged. The notation, ?$\|$, indicates which pairs of registers are tested for equality in order to end the phase. The flowchart is shown in Fig. 1-47.

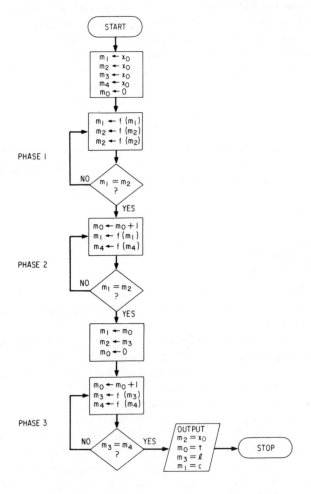

Fig. 1-47 Flowchart for inputting x_0 and outputting c, t, and l

Table 1-1. Use of Registers

Memory	Phase 1	Phase 2	Phase 3
m_0	$0 \; \text{-----} \; 0$	$\xrightarrow{\text{Increment}} c \quad ;$	$0 \xrightarrow{\text{Increment}} t$
m_1	$x_0 \xrightarrow{\;f\;} x_d$?‖	$\xrightarrow{\;f\;} x_{d+c} \quad ;$	$c \; \text{-----} \; c$
m_2	$x_0 \xrightarrow{\;ff\;} x_{2d}$	$\text{-----} \; x_d$?‖ $;$	$x_0 \; \text{-----} \; x_0$
m_3	$x_0 \; \text{-----} \; x_0$	$\text{-----} \; x_0$	$\xrightarrow{\;f\;} x_t$?‖
m_4	$x_0 \; \text{-----} \; x_0$	$\longrightarrow x_c$	$\xrightarrow{\;f\;} x_{t+c}$
No. of iterations→	d	c	t

Notice that in this program the function f appears seven times. (See Sec. 1.7 for a method of avoiding having to key it into the program seven times.)

Let us now analyze the time taken by this program. The function f is used three times in each of the d iterations in phase 1, twice in each of the c iterations in phase 2, and twice in each of the t iterations in phase 3, for a total of $3d + 2c + 2t$. Since $d < k = c + t$, the function f is used no more than 5k times. Each iteration of any of the three loops involves one comparison test, for a total of $d + c + t < 2k$ comparisons. The naive and Pollard methods are compared in Table 1-2.

The fact that the Pollard method uses only five, instead of $k + 3$, memories, proves its worth. The number of comparisons used, 2k, is much smaller than $1/2 \, k(k + 1) - t$, and there is also a great saving of time. Admittedly, function f gets used five times more in the Pollard ρ method than in the naive method, but this small disadvantage is greatly outweighed by the advantages.

We can see from this problem that cleverness can at times drastically improve the efficiency of a program. Some programs with a horrendously long running time, however, cannot be improved, no matter how clever the programmer is. That is, it has been *proved* that no improvement is possible. So how do you tell the difference between

Table 1-2. Comparison of Naive and Pollard Methods

	Naive Method	Pollard
No. of registers	$k + 3$	5
No. of uses of f	k	5k
No. of comparisons	$1/2 \, k(k + 1) - t$	2k

a program that can be made more efficient and one that cannot? This is the type of question for which there are no easy or universally applicable answers. Indeed, one of the most famous unanswered questions in computing theory is whether a class of problems called NP-complete can be solved by methods more efficient than the known techniques, which are so slow that the problems are for all practical purposes unsolvable. (See the Traveling Salesman Problem in Sec. 1.14.)

The ρ-diagram is representative of a function that eventually starts repeating itself when iterated. How many functions act this way? Some of them clearly never repeat, for example, the function $f(x) = x + 1$. On the other hand, any function whose rule can be programmed on a calculator *must* repeat eventually (that is, it must produce a ρ-diagram under iteration). The reason is that the calculator is capable of outputting only finitely many different numbers. If you program it to iterate a function, the program will necessarily repeat some output in the long run, and once this happens, you have entered the circular part of the ρ-diagram. There is a seeming paradox here, because the calculator can clearly be programmed to compute $x + 1$. Can you see the resolution?[34]

---

[34] See Sec. 1.8 for the answer.

Take number from all things and all things perish. Take calcu-
lation from the world and all is enveloped in dark ignorance
. . .

ISIDORE OF SEVILLE

CHAPTER TWO

Numbers

2.1 Introduction

Although your calculator belongs to a new technology, the language it speaks is very old. Virtually every society in history has developed at least a rudimentary number system. In fact, the oldest mathematical relic we possess is an animal bone inscribed with uniform scratches bundled in groups of five by cross-scratches the way we still do when tallying things. The scratches obviously represent numbers, and the bone is a number storage and retrieval device—from about 30,000 B.C.!

Where do numbers come from? They seem to be a spontaneous creation of the human mind or of the world. Whatever their origins, civilized man, starting from the "natural numbers"—1, 2, 3,. . . —has elaborated a language of incredible richness and fecundity for the purpose of investigating nature. Somehow the numbers retain a life of their own in spite of the many uses they are put to.

The study of numbers in their own right is called *number theory*. The oldest and one of the profoundest branches of mathematics, it continues to attract some of the most talented mathematicians to the present day. Almost every mathematician has tried his or her hand at solving at least one of the many unsolved problems in number theory—problems that are so easy to state and at the same time so intractable to solve (see Sec. 2.3 for an example).

This chapter contains a healthy dose of number theory. In problems such as Ulam's Problem, Pythagorean Triples, Sums of Digits, and Sums of Squares, the calculator executes algorithms that we could no more than *reason* about until recently because the computations involved are too laborious by hand. Other problems in the chapter, such as Multiple Precision and Complex Arithmetic, are of a more practical nature, but almost all are focused on numbers themselves. In Chap. 3, we will *use* numbers to model games.

2.2   The Universal Converter

Difficulty: 1

Now that the United States is changing over to the metric system, all of us are concerned with the problem of converting from one set of units to another. Several calculators on the market today have buttons for doing specific metric conversions—for example, from meters to feet—but in our opinion this is an extravagant waste of a calculator button. All a meters-to-feet button does is multiply the contents of the display by (approximately) 3.28. In fact, most unit conversions can be accomplished by multiplying by the appropriate constant.

Problem: Write a program that will do all such conversions in both directions. The constant for converting from the first system to the second should be stored in a memory.[1] The program should then accept numbers representing measurements in either system and convert to the other.

Solution: To convert from system A to system B; [see flowchart in Fig. 2-1(a)], you multiply by the constant stored in m_0. Can this same constant be used to convert from B to A? To answer this question, imagine that you started with a number representing a measurement

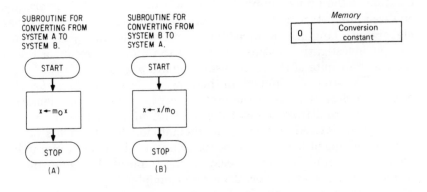

Fig. 2-1   Conversion flowcharts

---

[1] See the notes for how you come up with this number.

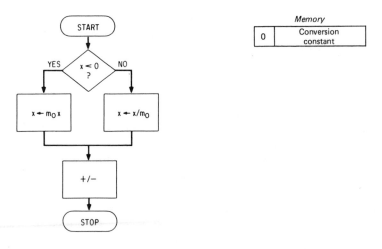

Fig. 2-2 Conversion flowchart using sign of input as cue

in system A and multiplied by the contents of m_0 to get the equivalent measurement in system B. How would you recover the original measurement? You would divide by the contents of m_0 to cancel the effect of the previous multiplication. Now imagine that all measurements in system B were arrived at by conversion from system A. You will then see that the equivalent measurement in system A is obtained by dividing by the contents of m_0. This is a simple example of a type of reverse thinking frequently used to solve mathematical problems: You are in state B and want to get to state A. You imagine a process that will get you from A to B, and then you reverse it to get the desired process in the other direction.

If your calculator has subroutines, the simplest thing to do is to write the two conversion procedures as two distinct subroutines. The user then simply calls the appropriate subroutine for the desired conversion [see Fig. 2-1(a) and (b)].

If your calculator doesn't allow for subroutines, you will have to construct your program in such a way that the calculator can tell which conversion you want. A nice technique is to use the sign of the input as a cue. Let's designate system A as the "negative" unit and system B as the "positive." Numbers preceded by a minus sign will then be interpreted as measurements in system A to be converted into system B, and vice versa. Let the program start with a test to see if the number in the display is less than 0. If so, multiply the contents of the display by the contents of m_0; if not, divide the contents of the display by m_0. As a final touch, change the sign of the result, cuing the user that the number has been converted into the other system. The flowchart is shown in Fig. 2-2.

This program is preferable to the two-subroutines approach since it is easier to think of one system of units being positive and the other negative than it is to remember the two conversions, SBR 1 and SBR 2 or GSB 01 and GSB 04. [*Solution was realized on an HP 33E in nine steps.*]

Notes: The latter program does have the disadvantage of being inoperative with negative units since the minus sign will be misinterpreted by the program as a cue. As a rule, however, this will not prove to be a serious drawback; one can always put the correct sign at the end. If this simple device doesn't work, the unit conversion is probably not of the type that the program is designed to handle. A case in point is the problem of converting from Fahrenheit to Celsius degrees. Clearly, -10 Fahrenheit is not the simple mirror image of $+10$. In fact, there is no natural connection between these two readings. You are probably also aware that the conversion formula from Fahrenheit to Celsius involves more than just multiplication by a constant.

How do you recognize the possibility of converting from one system to another by multiplying by a constant? An important requirement is that 0 must mean the same thing in both systems (imagine what happens when you convert). This requirement rules out Fahrenheit-Celsius conversion. The exact requirement is that there must be a constant ratio between measurements in one system and their equivalents in the other. The constant ratio is, in fact, the conversion constant. You can determine whether two systems of measurement satisfy this condition by looking at a table of equivalent values. Divide the entries on one side by their equivalents on the other; the answer should always be the same. For a concrete example, get out your camera and look at the ASA and DIN film-speed scales on it. Divide the ASA readings by the equivalent DIN readings, and you will find no constant ratio between them. Thus, ASA-DIN conversion is not accomplished by multiplying by a constant. On the other hand, there *is* a constant ratio between feet and meter readings on the lens, and this ratio is the conversion constant.

How do you find the conversion constant without a table of equivalent values? You can do so but not without coming up with at least one pair of equivalent readings. This can be a little tricky in the case of compound units. To show you how to go about it, suppose that you want to convert feet-per-second to kilometers-per-hour. The trick is to reason your way through via an intermediate unit. There are (approximately) 3281 feet in a kilometer. Thus, 1 kilometer-per-hour is equivalent to 3281 feet-per-hour. Since a second is 1/3600th of an hour, 1 foot-per-hour equals 1/3600 foot-per-second. One kilometer-per-hour, then, equals 0.9114 foot-per-second (3281/3600).

Here's a problem that you will encounter on your tire gauge before long. Find the conversion constant between pounds-per-square-inch and kilograms-per-square-centimeter. (Hint: There are 28.4 grams in an ounce and 3.28 feet in a meter).

Sometimes one can come up with a constant ratio between two systems after some adjustment of the numbers. Looking for such a connection can lead to the correct conversion formula. If, for instance, you subtract 32 from a Fahrenheit reading before comparing it with the corresponding Celsius reading (thus adjusting the zeros to the same place), you will find a constant ratio, suggesting that Fahrenheit-to-Celsius conversion is accomplished by first subtracting 32 from the Fahrenheit reading and then multiplying by a constant. Common connections to look for are ratios between entries on one scale and powers, roots, logs, and exponentials (antilogs) of entries on the other scale. (One of these connections works for ASA-DIN conversion.)

Another type of conversion problem not yet considered occurs when fractions of a given unit are expressed in an altogether different unit that is not a simple decimal subdivision of the larger unit. Two examples are degrees/minutes/seconds and feet/inches. Your calculator has keys for converting back and forth between decimal degrees and degrees/minutes/seconds (hours/minutes/seconds on some calculators). Can you construct an analogous conversion program for feet/inches?

2.3   Ulam's Problem

Difficulty: 1

Ulam's problem was posed by the Polish mathematician, Stanislaw Ulam. Consider the following algorithm. Let n be a positive whole number. If it is even, divide it by 2; if odd, multiply it by 3 and add 1. Applying the algorithm to the number 10 yields the number 5; applying it to 7 yields 22. The algorithm can be *iterated* (that is, applied to its own output). Starting from some number, iterating the algorithm produces a sequence of numbers. For example, starting with 10, successive iterations of the algorithm yield the sequence, $10 \rightarrow 5 \rightarrow 16 \rightarrow 8 \rightarrow 4 \rightarrow 2 \rightarrow 1$. Now when the sequence reaches 1, it gets caught in a loop: $1 \rightarrow 4 \rightarrow 2 \rightarrow 1 \rightarrow 4 \rightarrow 2 \rightarrow 1 \rightarrow \ldots$.

Ulam's problem is this: If you start with an arbitrary number n and iterate the stated algorithm, will it always crash into the $4 \rightarrow 2 \rightarrow 1$ loop? The answer is unknown. No one has ever produced a number that did not eventually crash. On the other hand, no one has been able to prove that every number must eventually crash. Even fairly small numbers can go a long way before crashing. Starting from 27, for example, 111 iterations of the algorithm are required before 1 is reached.

Problem: Write a program that takes as input a positive whole number n. Have the calculator flash the sequence of numbers obtained from n by successive iterations of Ulam's algorithm. In addition, put a counter in the program to keep track of the number of iterations. When (and if!) the sequence reaches 1, the program should halt and show the number of iterations it took to get to 1. Thus, for example, starting from 7, the calculator should output the sequence, 22, 11, 34, 17, 52, 26, 13, 40, 20, 10, 5, 16, 8, 4, 2, 1—followed by 15 (the number of iterations). Who knows? Maybe you will discover a way to show that Ulam's problem has an answer and become famous.

Solution: Basically, the program will apply the algorithm to the number n_0 to get a new number n_1, add 1 to the counter, and flash n_1. Check to see if $n_1 = 1$. If not, substitute n_1 for n_0 and start over; if so, recall the contents of the counter and stop.

Three memories are required, as follows:

1. m_0: The number to which the algorithm will be applied
2. m_1: The number obtained by applying the algorithm to the contents of m_0
3. m_2: The counter

The program (see the flowchart in Fig. 2-3) starts with some number in the display (either the first number in the sequence or the output from the previous iteration). Store this number in m_0, divide it by 2, and store the result in m_1. If the original number was even, m_1 will now contain the number obtained by applying Ulam's algorithm to it. Check this out by taking the fractional part of the number in the display (identical to the contents of m_1) and testing to see if it is 0. If it is, everything is OK and you should skip to the next paragraph. If not, recall m_0, multiply by 3, add 1, and store the result in m_1, which now contains the correct number.

Add 1 to the counter (m_2), recall m_1, and pause for one beat. Test to see if $m_1 = 1$ (you should be able to use the 1 that was added to m_2). If it isn't, hold the contents of m_1 in the display and loop back to the beginning of the program. If it is, recall the contents of m_2, clear m_2 for the next run, and halt. [*Solution was realized on an HP-25 in 22 steps.*]

Notes: The above program can be slicked up a little by using m_0 instead of m_1 as temporary storage for the new output of the algorithm. Begin by dividing the contents of m_0 by 2 and leave the result in m_0. Test to see if the result is a whole number. If it is, you're finished; if not, you've done the wrong part of the algorithm. Undo the error and produce the correct result in one blow by multiplying the contents of

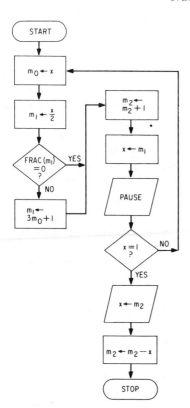

Memory	
0	Old n
1	New n
2	Counter

Fig. 2-3 Flowchart for Ulam's problem

m_0 by 6 and adding 1. Register arithmetic can be used to advantage here.

It seems strange to some people that mathematics should still present unsolved problems. After all (the argument runs), mathematical techniques are so mechanical, why can't machinery be set up for proving all possible mathematical truths? The answer is that mere machinery is not up to the job. Although mathematical proofs are indeed very mechanical once written down, the problem of assembling the machinery for proving a given theorem can not only require extraordinary human ingenuity, it sometimes exceeds it. Mathematics is and always has been a human endeavor. The history of mathematics shows how much intellectual and social conditions influence the quality and quantity of mathematics produced by a given culture in a given era.

In the past, there were hopes that a sufficiently clever human being might invent the ultimate theorem-proving machine, one capable of analyzing any mathematical statement and determining within a finite length of time its truth or falsehood. The problem was seriously considered in the seventeenth century by Leibniz, one of the inventors of calculus. Leibniz thought that not only mathematics, but in fact all knowledge, might be schematized and reduced to a mechanized treatment. Hilbert, in a famous speech to the First International Congress of Mathematicians in 1900, posed for the consideration of twentieth-century mathematicians the problem of constructing this ultimate machine, at least for mathematical purposes. All hopes for such a machine were dashed in 1930, however, when Kurt Gödel, a young German mathematician, proved its impossibility. Using a variation of the *liar paradox* ("Is the following sentence true or false?: 'This sentence is false.' "), Gödel showed that there are statements in number theory that are true in the sense that they have no counterexamples but are nevertheless not provable from the axioms. It is possible that Ulam's question is of this type. If so, of course, we will never know for sure because if we knew there were no counterexamples, we would have proved the answer to be "yes."

If counterexamples to Ulam's problem exist, they would be of two types. Either the number n would run into some cycle other than the 4, 2, 1 cycle or the sequence of iterates of n would "diverge to infinity." The first type would be verifiable on a calculator using the Pollard ρ algorithm (see Sec. 1.16). The second would not because the numbers in the sequence would eventually overflow the calculator's capacity, and one would have no way of knowing if they were eventually going to crash or not.

2.4   Phi, Fibonacci, and the World's Shortest Program

Difficulty: 1

Consider the following sequence: 1, 1, 2, 3, 5, 8, 13. Can you guess the next term? (Take your time; the problem can wait.) Since each term of the sequence is the sum of the previous two, the answer is 21. This innocent-looking string of numbers—the famous Fibonacci sequence—has inspired a mass of mathematical research. There is even a whole journal devoted to it, the *Fibonacci Quarterly*.

Problem: Write the shortest possible program for generating successive terms of the Fibonacci sequence. With an HP calculator, you should need no more than six steps. With a TI—depending on what type you have—you should need between seven (TI-57) and nine (TI-58) steps.

An amazing fact is that if you have an HP calculator, a program for this sequence exists that requires only one step! Actually, this is not quite true since all programs have to have a step to stop execution. On an HP-25 or 33E, this step is usually GTO 00, which not only stops execution but sets the program pointer to 01 to ready the program for the next run. If a program is going to accomplish anything, therefore, it must have at least two steps. Ignoring this control step, however, our statement *is* true. With the same proviso, there is a TI-57 program requiring three steps. Other TI programs will be slightly longer (the totals cited in the previous paragraph include a "stop" step).

How are these short programs capable of being realized? The answer lies in one of the curious properties of the Fibonacci sequence. Consider the ratios of its successive terms: 1/1, 2/1, 3/2, 5/3, 8/5, 13/8, ... These ratios converge to the "golden ratio" $[(\sqrt{5} + 1)/2]$, which is denoted by the Greek letter ϕ (phi).[2] Thus, if x_n and x_{n+1} are the nth and $n+1^{th}$ term of the sequence, x_{n+1}/x_n is approximately ϕ. Multiplying both sides of the approximation by x_n, we find that x_{n+1} is approximately ϕx_n. Each successive term of the sequence, then, can be obtained (approximately) by multiplying the previous term by ϕ, and the approximation gets better, not worse, as time goes on.

In order to get this approximate formula to turn out the precise terms of the Fibonacci sequence, two tricks are needed. First, the display must be set to zero decimal places, thereby rounding each term off to the nearest whole number and getting rid of most of the error. The second trick is to start in the right place. If you start at 1, successive multiplications by ϕ will almost generate the correct Fibonacci sequence, but not quite. The cumulative approximation errors deflect the sequence from the proper values. One should start instead with a number that absorbs all these errors from the outset, allowing the sequence to move *toward* the correct values gradually. As luck would have it, the early terms of the sequence will then be close enough to the correct terms if the calculator is set to zero decimal places.

The number to start with is $1/\sqrt{5}$.[3] [Note that $(1/\sqrt{5})\phi = .7236$, which is 1 when rounded off to zero decimal places, and that $(1/\sqrt{5})\phi^2 = 1.171$, which is also 1 when so rounded off.] You may verify on your own that successive multiplications by ϕ yield better and better approximations to the correct terms of the Fibonacci sequence. We now leave it to your use of the above information to find the shortest program for your calculator.

Solution: First, let's look at the "naive" solution. Each term of the sequence is the sum of the previous two. If we denote the terms by x_0,

---

[2] For a description of the golden ratio, see the Notes in Sec. 3.15.

[3] We will not discuss why this is the right number. The interested reader should look up Binet's formula.

$x_1, x_2, x_3, \ldots$, then the defining property for the Fibonacci sequence is given by the formula, $x_n = x_{n-2} + x_{n-1}$. The nth term is the sum of the $(n-1)^{st}$ and $(n-2)^{nd}$ terms. A straightforward program for generating the terms of the sequence will always keep the previous two terms of the sequence, x_{n-2} and x_{n-1}, stored in two memories, m_0 and m_1. To generate the next term, it will add the contents of m_0 and m_1. It will then store the result in m_1, moving the previous contents of the latter into m_0, ready for the next run, and leaving the new contents in the display. The HP and TI programs for this solution are as follows:

HP		TI	
01	RCL 0	00	RCL 0
02	RCL 1	01	+
03	STO 0	02	RCL 1
04	+	03	STO 0
05	STO 1	04	=
06	GTO 00	05	STO 1
		06	R/S

In the sophisticated solution, the program will start with $1/\sqrt{5}$ in the display, and each run will simply multiply the contents of the display by ϕ. Since the HP program will contain just one step— ☒ (or multiply)—where should ϕ be? The trick is to load the stack with ϕ's. Since the previous contents at the top of the stack will be left there as the stack drops, these ϕ's will replenish themselves automatically. At the beginning, then, the stack should look like the following:

T	ϕ
z	ϕ
y	ϕ
x	$1/\sqrt{5}$

Press ☒ once, and the stack will look like this:

T	ϕ
z	ϕ
y	ϕ
x	$(1/\sqrt{5})\phi$

Press ☒ again, and the stack will look like this:

T	ϕ
z	ϕ
y	ϕ
x	$(1/\sqrt{5})\phi^2$

and so forth. With the display set to zero decimal places, the correct terms of the Fibonacci sequence will appear. Since TI calculators don't provide a stack, a memory, say m_0, must be used to store ϕ. The TI-57 program (starting with $1/\sqrt{5}$ in the display) will then be the following:

```
00   ×
01   RCL 0
02   =
(03  R/S)
```

A little trickier, and more automatic, program would be the following:

```
00   ×
01   R/S
02   RCL 0
03   RST
```

This program need not be reset manually after each run. If you don't see why the ☐= is unnecessary, see Sec. 1.2(10).

Notes: Fibonacci was an early thirteenth-century mathematician from Pisa. Far ahead of his time, and considered by many historians to be the only European mathematician worthy of the name prior to the fourteenth century, he is credited with discovering the sequence that bears his name on the basis of a problem he posed in his book on arithmetic, *Liber Abaci*, written in 1202. (The book advocated the merits of the Arabic numeral system over the Roman numerals then in use.) The problem was formulated as follows:

How many pairs of rabbits, beginning with a single pair, will be produced in a year if in every month each pair bears a new pair that becomes productive from the second month on?

At the beginning, there will be one pair of rabbits. At the end of the first month, the pair will not yet be productive, and there will still be only one pair. At the end of the second month, the original pair will produce one new pair, and there will be two pairs. At the end of the third month, the original pair will produce yet another pair. The second pair will still not be productive, and the total number of pairs will be three. And so on until the end of the twelfth month. To express this sequence mathematically, suppose that x_n denotes the number of pairs of rabbits existing at the end of n months. Then x_n will be x_{n-1} plus the number of new pairs born in the nth month. This last number will be x_{n-2} because each pair that was alive two months previously will have produced a new pair during the nth month. Thus, $x_n = x_{n-1} + x_{n-2}$, and x_n is seen to be the nth term of the Fibonacci sequence. The answer to Fibonacci's question is therefore 233, the term x_{12} of the sequence.

Note that the one-step program given for HP calculators is a little silly. Since it requires you to press the $\boxed{\text{R/S}}$ key and all it does is press the $\boxed{\times}$ key, it would be just as efficient for you to press the $\boxed{\times}$ key yourself. The result is truly the shortest program possible: one with zero steps; that is, no program at all!

2.5   Sums of Digits

Difficulty: 2

An interesting function often occurring in number puzzles is the "sum of digits" function. Denoted by S(x), it takes any positive integer x (base 10) and computes the sum of its digits. Thus, S(340175) = 3 + 4 + 0 + 1 + 7 + 5 = 20; S(98) = 17; S(6) = 6; and so on.

Problem 1: Write a program that accepts as input a positive integer x and outputs S(x), the sum of the digits of x.

A curious property of the function S is that x and S(x) always leave the same remainder upon division by 9. Equivalently, x − S(x) is always evenly divisible by 9; for example, 340175 − S(340175) = 340175 − 20 = 340155 = 9 × 37795 and 98 − S(98) = 98 − 17 = 81 = 9 × 9.

Let us now consider the effect of iterating the function S on some positive integer x, that is, let us consider the sequence: x, S(x), S[S(x)], S(S[S(x)]), ... For example, if x = 157403431, we get the sequence: 157403431, 28, 10, 1, 1, 1, ...

For any positive integer x, S(x) ≤ x, that is, the sum of the digits of a number is never bigger than the number itself. Furthermore, S(x) < x in all cases except when x is a one-digit number. For any positive integer x, therefore, the sequence of x, S(x), S[S(x)], S(S[S(x)]), . . . will consist of smaller and smaller numbers until it reaches a one-digit number; thereafter, it merely repeats that number.

Since x and S(x) always leave the same remainder upon division by 9, it follows that every number in this sequence will leave the same remainder upon division by 9. In particular, when the sequence stabilizes at a one-digit number, r, that number *must* be the remainder upon dividing x by 9 (unless the repeating one-digit number is 9, when the remainder is 0.)

Problem 2: Write a program that accepts as input a positive integer x and calculates the remainder r upon dividing x by 9.

Solution to Problem 1: Think of input x not as a number, but as a row of separate digits all stored in one register. What we want to do is fish the digits out one at a time and sum them up. The technique for fishing them out is described in Sec. 1.14.

We will need two memories. Memory m_0 will initially contain x and then decrease as digits are fished out of it. Memory m_1 will initially contain 0 and then accumulate the sum of the digits $S(x)$, successively increasing as newly fished out digits are added to it. The flowchart is shown in Fig. 2 / 4.

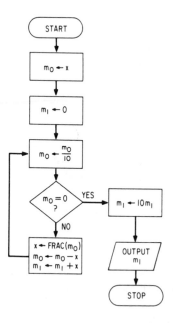

Fig. 2-4 Flowchart for inputting a positive integer x and outputting $S(x)$

Each time through the loop in this program, a new digit is fished out of m_0 and summed into m_1. The step, $m_0 \leftarrow m_0/10$, slides the decimal point one place to the left, pushing the rightmost digit of what's left in m_0 past the decimal point. The test, $m_0 = 0$, prevents the loop from going on forever; when $m_0 = 0$, all the digits have been fished out. If $m_0 \neq 0$, the loop continues. The step, $x \leftarrow \text{FRAC}(m_0)$, isolates the rightmost digit of m_0 (the one now on the right of the decimal point) from the rest, whereas the step, $m_0 - x$, removes this digit from m_0. The step, $m_1 \leftarrow m_1 + x$, sums the newly fished out digit into register m_1, and the program returns to the top of the loop.

Notice that the digits are fished out from the right side of the decimal point, that is, they are really one-tenth of what they should be. Hence, m_1 doesn't accumulate $S(x)$, but rather $(1/10)S(x)$. This accounts for the step, $m_1 \leftarrow 10m_1$, before outputting m_1.

Solution to Problem 2:   We will set up a loop to use the flowchart in Fig. 2–4 over and over again, beginning with an input of x and at each stage using the resulting output for the next input as shown in the flowchart of Fig. 2–5.

The steps inside the broken-line rectangle are from Fig. 2–4. The step, $m_0 \leftarrow 10m_1$, takes the output $10m_1$ of this rectangle and places it back in input m_0. Before returning, it tests to see if it is done, that is, whether or not the output now in m_0 is a one-digit number. If so (if $m_0 \leq 9$), the program is almost done; m_0 is the right answer r, except for the case, $m_0 = 9$, when the right anwer is zero instead. This exception accounts for the test, $m_0 = 9$?, and the steps following it. Otherwise (if $m_0 \neq 9$), m_0 is at least a two-digit number, and the program goes to the top of the loop.

Notes: Although the problems in this section are of little relevance today, they were quite important back in the days of hand computation because they provided a simple test for checking answers and detecting careless errors. The procedure was called "casting out nines." It is not infallible, but it does catch about 8 out of 9 errors. Let's look at a brief example and consider the sum

$$
\begin{array}{r}
19815 \\
+\ 23288 \\
\hline
=\ 43203
\end{array}
$$

Is the anwer correct? To check, we first calculate the r-value of the summands 19815 and 23288. For 19815, r = 6, and for 23288, r = 5. The sum of these two r-values is 11, and the r-value of 11 is 2. Here is the test: If the sum is correct, the r-value of the answer should also be 2, but it is not. For 43203, r = 3. The sum is therefore incorrect. It should be 43103, for which r = 2.

There are rules for "casting out n's" for any number n. For casting out 6's, the rule is: Multiply each digit, except the rightmost one, by 4, leaving the last digit alone; add up the results; then, when the sum is divided by 6, the remainder will be the same as when the original number is divided by 6. Let's try 5283 to illustrate. Using our rule, we form the sum: $(4 \times 5) + (4 \times 2) + (4 \times 8) + 3 = 63$. When the latter number is divided by 6, it leaves a remainder of 3, this remainder occurring when 5283 is divided by 6 (try it out). As in Problem 2, we can iterate this procedure until a one-digit number is reached. Applying it to 63, we get: $(4 \times 6) + 3 = 27$; applying it to 27, we get $(4 \times 2) + 7 = 15$; and applying it to 15: $(4 \times 1) + 5 = 9$. Now we can easily verify that division by 6 leaves a remainder of 3 (although not quite so easily as in the case of 9).

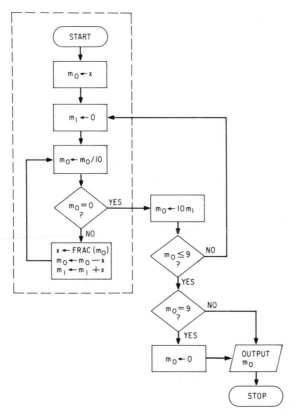

Fig. 2-5 Flowchart for inputting a positive integer x, dividing by 9, and outputting remainder r

The rule for casting out 3's is just like the rule for casting out 9's. As in the case of 6's, a last division may have to be performed to find the remainder; thus the program for Problem 2 can be modified to find remainders upon division by 3. (See Sec. 2.11 for another algorithm for finding remainders—one that is faster on a calculator but more difficult by hand.)

2.6   Sums of Squares

Difficulty: 3

The sequence, 0, 1, 4, 9, 16, 25, 36, 49, 64, 81, 100, . . ., etc., of (perfect) squares has inspired countless problems and puzzles (see Sec. 2.8). In this section, we will examine what happens when we add a few squares. Let's consider first the simplest question concerned with adding squares: Which numbers are the sum of two squares? The number 80, for example, is the sum of two squares: $80 = 16 + 64 = 4^2 + 8^2$. The number 3, on the other hand, is not. Is 21 the sum of two squares? Is 74?

Problem 1: Write a program that will accept as input any nonnegative integer m and that will output all pairs (x, y) of nonnegative integers such that $x \leq y$, and

$$x^2 + y^2 = m$$

If you solve Problem 1, use it to verify the following table:

m	Solutions (x, y)		Solutions (x, y)
0	(0,0)	20	(2,4)
1	(0,1)	21	
2	(1,1)	22	
3		23	
4	(0,2)	24	
5	(1,2)	25	(3,4),(0,5)
6		26	(1,5)
7		27	
8	(2,2)	28	
9	(0,3)	29	(2,5)
10	(1,3)	30	
11		31	
12		32	(4,4)
13	(2,3)	33	
14		34	(3,5)
15		35	
16	(0,4)	36	(0,6)
17	(1,4)	37	(1,6)
18	(3,3)		
19			

Notice that for some m, like 3, 6, 11, or 12, there are no solutions at all to the equation, $x^2 + y^2 = m$, whereas for others (try m = 32045), there are many. (If your calculator has a printer on it, perhaps you can program it to print out the above list.)

For an easier way to determine whether there are any solutions to the equation, $x^2 + y^2 = m$, see any text on elementary number theory. See also the historical note at the end of this section.

Now consider the question: Which numbers are sums of three squares?

Problem 2: Write a program that will accept as input a given nonnegative integer n and output all triples (x, y, z) of nonnegative integers such that $x \leq y \leq z$ and

$$x^2 + y^2 + z^2 = n$$

A list of solutions for small values of n (make your calculator prepare it if it has a printer) should look like this:

n	(x,y,z)		(x,y,z)
0	(0,0,0)	9	(0,0,3),(1,2,2)
1	(0,0,1)	10	(0,1,3)
2	(0,1,1)	11	(1,1,3)
3	(1,1,1)	12	(2,2,2)
4	(0,0,2)	13	(0,2,3)
5	(0,1,2)	14	(1,2,3)
6	(1,1,2)	15	
7		16	(0,0,4)
8	(0,2,2)		

Again, notice that certain numbers n, like 7 and 15, have no solution. In this case, it is easy to describe precisely which numbers n have no solution. If a and b are nonnegative integers, then a number n equal to $4^a(8b + 7)$ admits *no* solution to the equation, $n = x^2 + y^2 + z^2$. For example, when a = 0 and b = 0 or 1, we have the first two such numbers: 7 and 15. *All* other nonnegative integers n have a solution.

Problem 3:   Write a program that will accept as input any nonnegative integer p and output all nonnegative integers, w, x, y, z, such that $w \leq x \leq y \leq z$ and

$$w^2 + x^2 + y^2 + z^2 = p$$

To test your program, here is a list of all solutions for small values of n:

p	(w,x,y,z)	p	(w,x,y,z)
0	(0,0,0,0)	6	(0,1,1,2)
1	(0,0,0,1)	7	(1,1,1,2)
2	(0,0,1,1)	8	(0,0,2,2)
3	(0,1,1,1)	9	(0,0,0,3),(0,1,2,2)
4	(0,0,0,2),(1,1,1,1)	10	(0,0,1,3),(1,1,2,2)
5	(0,0,1,2)		

If you extend this list to, say, p = 100, you will discover that all 100 values of p have at least one solution in contrast to n and m where some values had no solution. Does this go on forever? Does *every* nonnegative integer p have a solution? You cannot settle this question by using you program for ever larger and larger values of p unless you happen to find some number p with no solution fairly early on. But the fact is, *all* values of p do have a solution! That is, every nonnegative integer is a sum of four squares (see the Notes).

Solution to Problem 1: Given m, we want to find all pairs (x, y) of nonnegative integers satisfying the equation, $x^2 + y^2 = m$, where $x \leq y$. If we knew y, we could easily find x since solving the equation gives

$$x = \sqrt{m - y^2}$$

Our plan is therefore to try all possible values of y, pausing to output the pair (x, y) *if* $\sqrt{m - y^2}$ happens to be an integer. (The test, $x = \text{INT } x$?, can be used to see whether or not x is an integer.

What do we mean by "all possible values of y"? Well, since $0 \leq x^2$ for any x, $y^2 \leq x^2 + y^2 = m$, or $y \leq \sqrt{m}$. Also, since $x \leq y$, obviously $x^2 \leq y^2$. Thus, $m = x^2 + y^2 \leq y^2 + y^2$, or $\sqrt{m/2} \leq y$. We therefore have to check only those integers y in the range $\sqrt{m/2} \leq y \leq \sqrt{m}$. We will use register m_0 to store these values of y successively, beginning with $\text{INT}(\sqrt{m})$—the largest integer in the range—and working down. Another register, m_3, will permanently store m. The flowchart is shown in Fig. 2-6.

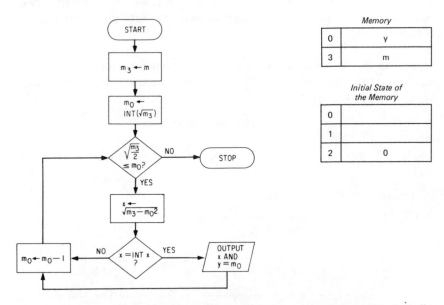

Fig. 2-6 Flowchart for inputting any nonnegative integer m and that will output all pairs (x,y) of nonnegative integers such that $x \leq y$ and $x^2 + y^2 = m$

In the test, $\sqrt{m_3/2} \leq m_0$, the quantity $\sqrt{m_3/2}$ can be computed directly each time through, or it can be calculated just once at the beginning of the program and stored permanently in another register, such as m_6.

Solution to Problem 2: Given n, we wish to find all triples (x, y, z) or nonnegative integers such that $x \leq y \leq z$ and

$$x^2 + y^2 + z^2 = n$$

If we knew z, we could find all the x's and y's by using Problem 1, letting $m = n - z^2$, to solve

$$x^2 + y^2 = n - z^2$$

Our plan, then, is to try all possible values of z, pausing to output solutions (x, y, z) as they are found.

What are "all possible values of z"? Well, since $0 \leq x^2 + y^2$, it is clear that $z^2 \leq x^2 + y^2 + z^2 = n$. Therefore, $z \leq \sqrt{n}$. Also, since $x \leq y \leq z$, it is clear that $x^2 \leq z^2$ and $y^2 \leq z^2$. Thus, $n = x^2 + y^2 + z^2 \leq z^2 + z^2 + z^2$, or $n/3 \leq z^2$. We therefore have to check only those integers z in the range, $\sqrt{n/3} \leq z \leq \sqrt{n}$. These values will be stored in memory m_1, beginning with the largest [$m_1 = \text{INT}(n)$] and decreasing to the smallest. We will use m_4 to store n permanently.

The flowchart is shown in Fig. 2–7. The steps inside the broken rectangle correspond to the flowchart in Fig. 2–6. The quantity $\sqrt{m_4/3}$, appearing in the test, $\sqrt{m_4/3} \leq m_1$?, can be calculated directly each time it is used, or it can be just calculated once, at the beginning of the program, and stored permanently in another register (like m_7, for example).

Solution to Problem 3: Given p, we wish to find all quadruples (x, y, z, w) of nonnegative integers such that $x \leq y \leq z \leq w$ and

$$x^2 + y^2 + z^2 + w^2 = p$$

If we knew w, we could find all x's, y's, and z's by using Problem 2, letting $n = p - w^2$, to solve

$$x^2 + y^2 + z^2 = p - w^2$$

Our plan is therefore to try all possible values of w, pausing to output solutions (x, y, z, w) as they are found.

To find the range of values of w that need to be tested, note that $0 \leq x^2 + y^2 + z^2$ and consequently that $w^2 \leq x^2 + y^2 + z^2 + w^2 = p$, or $w \leq \sqrt{p}$. Also, since $x \leq y \leq z \leq w$, we know that $x^2 \leq w^2$, that $y^2 \leq w^2$, and that $z^2 \leq w^2$. Thus, $p = x^2 + y^2 + z^2 + w^2 \leq w^2 + w^2 + w^2 + w^2 = 4w^2$ so that $\sqrt{p/4} \leq w$. As a result, we need check only those integers w in the range $\sqrt{p/4} \leq w \leq \sqrt{p}$. These values will be stored in register m_2, beginning with the largest [$m_2 = \text{INT}(\sqrt{p})$] and decreasing to the smallest. We will use m_5 to store p permanently.

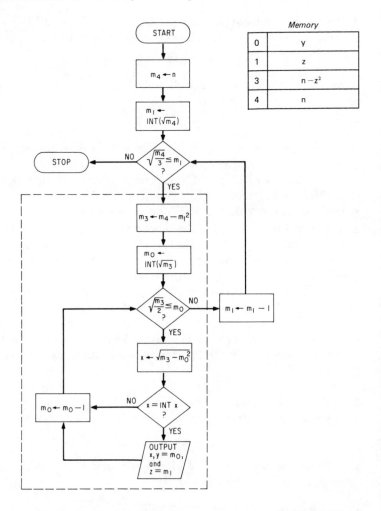

Fig. 2-7 Flowchart for inputting any nonnegative integer n and outputting all triples (x,y,z) of nonnegative integers such that $x \le y \le z$ and $x^2 + y^2 + 3^2 = n$

The flowchart is shown in Fig. 2-8. The steps inside the broken rectangle are essentially those in the flowchart of Fig. 2-7. The number $m_5/4$ can be calculated anew everytime it is used in the test, $\sqrt{m_5/4} \le m_2$?, or it can be calculated just once at the beginning of the program and permanently stored in some register, such as m_8.

Historical note: The theory of the sums of two squares is well known today, and using it one can answer such questions as "Which numbers

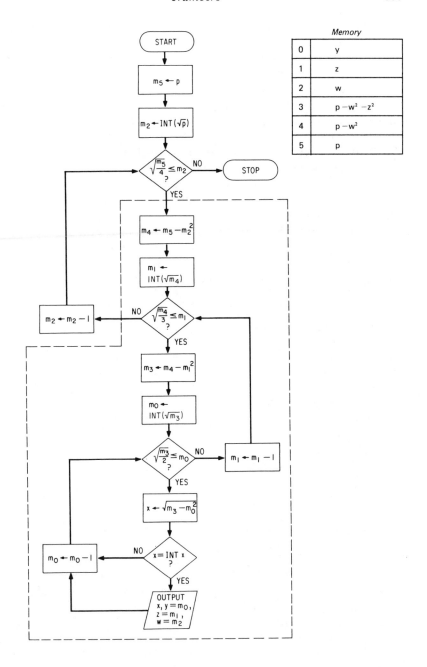

Fig. 2-8 Flowchart for finding all quadruples (x,y,z,w) of nonnegative integers such that $x \leqslant y \leqslant c6z \leqslant w$ and $x^2 + y^2 + 3^2 + w^2 = p$

are the sum of two squares?" and "In how many different ways is a given number the sum of two squares?" with relative ease. At the heart of the theory lies a theorem first expressed by Pierre de Fermat (1601–65), as follows: If p is a prime number of the form 4n + 1, then p is the sum of two squares in exactly one way.

Fermat, a lawyer by profession, claimed to practice mathematics only as an amateur and seldom published proofs of his work. The first published proof of his theorem came from Leonhard Euler (1707–83). Joseph Louis Lagrange (1736–1813) was the first to publish a proof of Fermat's assertion that every nonnegative integer is the sum of four squares. (Euler and Lagrange are considered the strongest mathematicians of the eighteenth century.) Later, Carl G. J. Jacobi (1804–51) developed a formula for the number of different ways that a given number can be expressed as a sum of four squares.

It is interesting to note that sums of three squares are harder to deal with than sums of two or four squares. The main stumbling block may be explained as follows. The formula,

$$(a^2 + b^2)(c^2 + d^2) = (ac + bd)^2 + (ad - bc)^2$$

shows that the product of any two numbers which are themselves the sum of two squares is again the sum of two squares. A similar formula holds for four squares; that is, the product of two numbers which are themselves the sum of four squares is again the sum of four squares:

$$(a_1^2 + b_1^2 + c_1^2 + d_1^2)(a_2^2 + b_2^2 + c_2^2 + d_2^2) =$$
$$(a_1a_2 - b_1b_2 - c_1c_2 - d_1d_2)^2 + (a_1b_2 + b_1a_2 + c_1d_2 - d_1c_2)^2 +$$
$$(a_1c_2 - b_1d_2 + c_1a_2 + d_1b_2)^2 + (a_1d_2 + b_1c_2 - c_1b_2 + d_1a_2)^2$$

However, no such formula holds for the sum of three squares! In fact, $3(= 1^2 + 1^2 + 1^2)$ and $5(= 0^2 + 1^2 + 2^2)$ are sums of three squares, but $3 \times 5(= 15)$ is not.

2.7   Recursions

Difficulty: 3

The Fibonacci numbers, as you may recall from Sec. 2.4, are a sequence $f_0, f_1, f_2, \ldots$, defined as follows: $f_0 = 0$, $f_1 = 1$, and $f_n = f_{n-1} + f_{n-2}$ for each $n \geq 2$. Notice that even though no formula is given here for computing f_n directly, we can still find each f_n by using only the information given above. An indirect formula of this kind is called a *recursion*. Many important sequences can be defined recursions, that is, by rules that allow you to calculate each number in the sequence in terms of previous ones.

Here are five examples. You should compute the first few terms of each sequence by hand and then check them with our answers to make sure you understand how recursions work:

(1) $a_0 = 0$, $a_1 = 1$, $a_2 = 4$, and $a_n = 3a_{n-1} - 3a_{n-2} + a_{n-3}$ for $n \geq 3$
(2) $b_0 = 1$ and $b_n = n(b_{n-1})$ for $n \geq 1$
(3) $c_0 = 1$, $c_1 = 2$, and $c_n = 4c_{n-1} - c_{n-2}$ for $n \geq 2$
(4) $d_0 = 0$, $d_1 = 1$, and $d_n = 4d_{n-1} - d_{n-2}$ for $n \geq 2$
(5) $e_0 = 0$, and for $n \geq 1$, if n is a prime number, the $e_n = 1 + e_{n-1}$; if n is not prime, then $e_n = e_{n-1}$ (a prime number is a number larger than one that is evenly divisible only by itself and 1)

Here is a table of the first few numbers in these sequences:

n	0	1	2	3	4	5	6	7	8
a_n	0	1	4	9	16	25	36	49	64
b_n	1	1	2	6	24	120	720	5040	40320
c_n	1	2	7	26	97	362	1351	5042	18817
d_n	0	1	4	15	56	209	780	2911	10864
e_n	0	0	1	2	2	3	3	4	4

The first sequence is nothing more than the sequence of squares: $a_n = n^2$. You might recognize the second sequence, $b_n = n!$, as the product of the numbers n, n − 1, n − 2, $\cdots$, 3, 2, 1. The third and fourth sequences are related. It turns out that for every n, $c_n^2 - 3d_n^2 = 1$. [Try it: $1^2 - (3 \cdot 0^2) = 1$; $2^2 - (3 \cdot 1^2) = 1$; $7^2 - (3 \cdot 4^2) = 1$; etc.] Furthermore, the sequences c_n and d_n contain *all* pairs of non-negative integers c and d satisfying the equation, $c^2 - 3d^2 = 1$. Finally, although the numbers in the fifth sequence are much smaller than the numbers in the other four, it is a much more complicated sequence, e_n being the number of prime numbers not greater than n.

Problem: Write four programs, one each for calculating the first four sequences above using the given recursions.

Solution: We will describe only the first sequence and let you find the others. Note that the formula,

$$a_n = 3a_{n-1} - 3a_{n-2} + a_{n-3}$$

requires three previous terms to calculate the subsequent one. Thus we need three memories to store the last three terms.

The flowchart is shown in Fig. 2–9. The first step calculates the next term, and the succeeding three steps update the three mem-

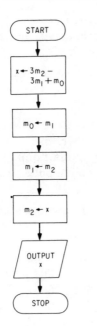

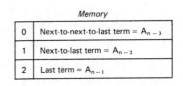

Fig. 2-9 Flowchart for a recursion

ories for the next term—before displaying the term x just calculated. Be careful that you don't lose x while you are shifting the memories around. (You might need to use another memory to store it temporarily.)

2.8   Pythagorean Triples

Difficulty: 2

Recall the famous Pythagorean theorem: If a, b, and c are the legs and hypotenuse on a right triangle (see Fig. 2-10), then $a^2 + b^2 = c^2$. If a, b, and c are (positive) *whole numbers* satisfying this equa-

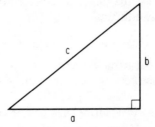

Fig. 2-10 Triangle to illustrate the Pythagorean theorem

tion, then a, b, and c are called a *Pythagorean triple*. The simplest example is the triple, 3, 4, and 5 ($3^2 + 4^2 = 9 + 16 = 25 = 5^2$). Pythagorean triples have fascinated mathematicians for millenia (see the Notes). There are lots of them—infinitely many to be precise—but they are rather thinly scattered. The next interesting one after 3, 4, and 5 is 5, 12, and 13 ($5^2 + 12^2 = 25 + 144 = 169 = 13^2$).[4]

Problem 1: Write a program that finds all Pythagorean triples.

If you manage to solve this problem and run your program, you will soon note how slow it is since the larger the numbers in the triples get, the longer it takes the calculator to find them. You can imagine, then, how hard it would be to find Pythagorean triples unaided by any device or method. A long time ago certain clever individuals began looking for formulas that would generate Pythagorean triples automatically.

Here is a scheme attributed to Pythagoras himself. Pythagoras noticed that the sum of all odd numbers up to any given point always add up to a perfect square; for example, $1 + 3 = 4 = 2^2, 1 + 3 + 5 = 9 = 3^2, 1 + 3 + 5 + 7 = 16 = 4^2$, etc. Now suppose that the last term in such a sum is itself a perfect square, as in the sum, $1 + 3 + 5 + 7 + 9$. Then the whole sum is not only a perfect square (25 in this case) but can be broken into two other perfect squares (namely, the sum up to, but not including, the last term and the last term by itself), thereby creating a Pythagorean triple in this case, $25 = 1 + 2 + 3 + 5 + 7 + 9 = (1 + 3 + 5 + 7) + (9) = 16 + 9 = 4^2 + 3^2$. The next sum for which this trick works is $1 + 3 + 5 + \ldots + 21 + 23 + 25$ since $13^2 = 169 = 1 + 3 + 5 + \ldots + 21 + 23 + 25 = (1 + 3 + 5 + \ldots + 21 + 23) + (25) = 144 + 25 = 12^2 + 5^2$, yielding the Pythagorean triple, 5, 12, and 13.

It becomes clear that *any* odd perfect square, m^2, sits at the end of a sum of consecutive odd numbers, $1 + 3 + 5 + \ldots + (m^2 - 2) + m^2$, on which we can perform this trick. We will spare you the details that show that $1 + 3 + 5 + \ldots + (m^2 - 2)$ adds up to $[(m^2 - 1)/2]^2$ and that the whole sum, $1 + 3 + 5 + \ldots + (m^2 - 2) + m^2$, adds up to $[(m^2 + 1)/2]^2$, producing the Pythagoren triple, m, $(m^2 - 1)/2$, and $(m^2 + 1)/2$. Thus we have Pythagoras' formula: If m is any odd number, then m, $(m^2 - 1)/2$, and $(m^2 + 1)/2$ are a Pythagorean triple.

Problem 2: Write a program that generates Pythagorean triples by using Pythagoras' formula (no solution is provided).

---

[4] An uninteresting one is 6, 8, and 10, which is just the 3, 4, and 5 triple enlarged by a factor of two.

Pythagoras' scheme does not generate *all* Pythagorean triples. If you solve Problem 2, you will probably notice that the program produces only those triples a, b, and c where c = b + 1 [that is, ½(m² + 1) = ½(m² − 1) + 1]. Thus it will not generate the triple, 8, 15, and 17, which you can readily verify to be Pythagorean. What is needed to produce all triples is a formula taking two numbers as input instead of just one.

Let z and w be any two positive whole numbers with w < z. Then a, b, and c where a = z² − w², b = 2zw, and c = z² + w² are a Pythagorean triple since a² + b² = (z² − w²)² + (2zw)² = z⁴ − 2z²w² + w⁴ + 4z²w² = z⁴ + 2z²w² + w⁴ = (z² + w²)² = c². For example, let z = 7 and w = 4. Then a = 49 − 16 = 33, b = 2 × 7 × 4 = 56, and c = 49 + 16 = 65. You can check on your own that 33, 56, and 65 are a Pythagorean triple.

This scheme generates all triples except for a few uninteresting ones. (It does not, for instance, generate the triple, 9, 12, and 15, which is just the triple, 3, 4, and 5 enlarged by a factor of 3.)[5] It *does* generate all "primitive" ones (that is, those that are not multiples of others).

Problem 3:   Write a program for generating all Pythagorean triples that makes use of the preceding formulas.

Solutions to Problems

Problem 1:   Since a² + b² = c² and c = $\sqrt{a^2 + b^2}$, the Pythagorean triple, a, b, and c, can be rewritten as a, b, and $\sqrt{a^2 + b^2}$. We thus generate all triples, a, b, and $\sqrt{a^2 + b^2}$, checking each time to see if $\sqrt{a^2 + b^2}$ is a whole number. If it is, then a, b, and $\sqrt{a^2 + b^2}$ are a *Pythagorean* triple, and the program stops to output it. We may as well generate only those triples for which a ≤ b since the others are redundant (the triple, 4, 3, and 5, is no different from the triple, 3, 4, and 5). Thus we need part of the program to generate all possible pairs of whole numbers a, b such that a ≤ b. This we accomplish with nested loops (see Sec. 1.8), as shown in the flowchart of Fig. 2-11. Now all the rest of the program has to do is compute $\sqrt{a^2 + b^2}$, store it in a memory (should it be a whole number), and check to see if it is a whole number (fractional part = 0?). If it is, the program then outputs the triple, a, b, and c. We will leave it to you to arrange the output (see

---

[5] Unfortunately, it does not avoid *all* uninteresting triples. Note this exception, among others: If z = 3 and w = 1, we get the uninteresting triple, 6, 8, and 10. See the Notes for more on the subject.

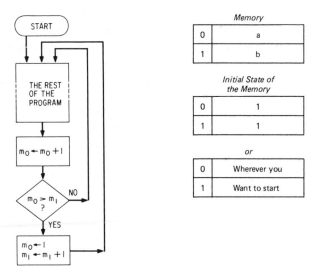

Fig. 2-11 Flowchart for finding all Pythagorean triples (first part)

Chap. 1). If $\sqrt{a^2 + b^2}$ is not a whole number, it is time to generate a new pair, a, b, which returns to the part of the program already described. The flowchart for the remainder of the program is shown in Fig. 2-12.

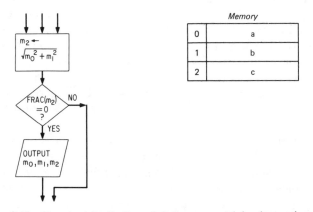

Fig. 2-12 Flowchart for finding all Pythagorean triples (second part)

Problem 3: In this program we want to generate all pairs, z, w such that w < z and to generate from each pair the triple, a, b, and c, using the formulas given. The part of the program generating the pairs z, w will be just like that for generating the pairs a, b in the previous solution, except that the test, $m_0 > m_1$?, is replaced by the test, $m_0 = m_1$? The flowchart for the rest of the solution is shown in Fig. 2-13. (Register arithmetic can be used to advantage in this program.)

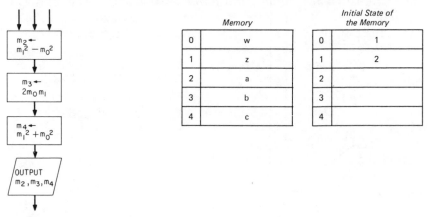

Fig. 2-13 Flowchart for generating all pairs (z,w) such that w ≤ z and from each pair the triple (a,b,c)

Notes: Pythagoras got his name attached to the Pythagorean theorem and Pythagorean triples more or less by historical accident. Until the twentieth century, it had been thought that there was very little mathematics worth mentioning before his time. However, we now know that the Egyptians and Babylonians were producing respectable mathematics as far back as 2000 B.C., 1400 years before he lived. Indeed, the more we learn about their work (especially that of the Babylonians) the more impressive it becomes. Clay tablets dating from about 1800 B.C. show that Babylonians already knew how to generate Pythagorean triples. Professor Otto Neugebaur has argued, based on the evidence of the tablets (which are lists of Pythagorean triples and related numbers), that the Babylonians had even discovered the general formula used in Problem 3.

Nevertheless, Pythagoras well deserves his eminent place in the history of mathematics. Born around 600 B.C., he is one of the great patriarchs of science although his actual life is shrouded in legend. Since the oldest biography of him was written approximately 600 hundred years after his death by a man (Iamblichus) who clearly regarded him as a divine figure, virtually all that is now "known" of his life must be viewed with some skepticism.

The story goes that after spending the early part of his life traveling and studying in Egypt and Babylonia, Pythagoras moved from his native Samos (an island off the coast of modern Turkey) to southern Italy, where he founded a school. His school was devoted to the study of mathematics, its motto being "All is number." By this, Pythagoras meant that all things are knowable through numbers. Basically, this is still the view of modern science for science seeks to know the world through mathematical formulas, and such formulas speak in numbers. It is for this reason that Pythagoras is credited with being the primogenitor of the scientific worldview. Yet Pythagoras' own conception of the meaning of "All is number" was much broader than the modern scientist's. He seems to have thought that *literally every- thing*—man, love, justice, beauty—was explainable by numbers.

Pythagoras discovered the numerical ratios that explain the construction of the musical scale. He thought the harmonies that the ear perceives in music to be material manifestations of a higher harmony present in the numbers that govern the music. And he may have extended this view to the whole universe, believing that all of physical nature was merely an "imitation" of the world of numbers. His followers, at any rate, devoted themselves to the contemplation of the harmonies of numbers available to the mind alone. Pythagoras' conception influenced Plato, whose philosophy replaced numbers by "forms."

As we noted earlier, the scheme for generating Pythagorean triples in Problem 3 produces all the "primitive" triples, those from which all others can be derived by multiplication. From the primitive triple, 5, 12, and 13, for example, we can generate the triples, 10, 24, and 26 (multiply by 2); 15, 36, and 39 (multiply by 3); etc. But 5, 12, and 13 are not themselves a multiple of any other triple. Clearly, if we are interested in generating *all* Pythagorean triples, it suffices to generate *only* the primitive ones—a job that the scheme in Problem 3 doesn't do. In addition to the primitive triples, it generates some (though not all) nonprimitive ones (see the footnote). The nonprimitive triples can be avoided by imposing two restrictions on the numbers z and w:

1. They should have no common divisor other than 1 (that is, they should be *relatively prime*).
2. They should not both be odd (they cannot both be even either, since 2 would then be a common divisor).

It is tricky to write a program embodying these restrictions; we leave it to the fanatics among our readers.

Pythagorean triples can be generalized in various ways. There are, for example, "Pythagorean quadruples," three perfect squares adding up to a perfect square, such as 6^2, 10^2, 15^2, and 19^2. One can also find quadruples of perfect cubes ($3^3 + 4^3 + 5^3 = 6^3$), quintuples of perfect fourth powers ($30^4 + 120^4 + 272^4 + 315^4 = 353^4$), sextuples of perfect fifth powers, and so forth. However, there are no *pairs* of cubes adding up to a perfect cube, that is, there are no triples of (positive) whole numbers a, b, and c such that $a^3 + b^3 = c^3$. An incomplete proof of this fact was given by Euler in the eighteenth century. In the nineteenth century, Gauss, perhaps the greatest mathematician of all time, gave the first entirely correct proof.

Gauss' proof was a special case of a theorem proposed by the seventeenth-century number theorist, Pierre de Fermat. Fermat wrote in the margin of one of his algebra books, next to a discussion of Pythagorean triples, that for *all* whole numbers n larger than 2, the equation, $a^n + b^n = c^n$, has no whole number solutions. He claimed that he had an elegant proof of this fact, but that the margin was too small to contain it. The note was discovered after Fermat's death and the proof has never been found. It is now referred to as Fermat's Last Theorem (or Lost Theorem) and is probably the most famous unsolved problem in all of mathematics.

2.9   Permutations and Combinations

Difficulty: 2

In how many ways can Alice, Bill, and Charlie line up for a picture? A little reflection will show that there are six ways, represented by the three-letter "words" ABC, ACB, BAC, BCA, CAB, and CBA. Suppose now that Dan joins them. In how many ways can the four of them line up? It turns out that there are 24 ways, a figure that you should check. It is, of course, the same as the number of four letter "words" that can be made by rearranging ABCD.

The number of ways n people can line up for a picture, or the number of n-letter "words" that can be made by rearranging n different letters, is denoted by the symbol n! (read "n factorial"). From the examples above, 3! = 6 and 4! = 24. There is a formula for n!, as follows:

$$n! = n(n - 1)(n - 2) \ldots (3)(2)(1)$$

That is, n! is the product of all positive whole numbers up to and including n. Here is your first problem, an easy one for which we will give no solution (see Sec. 1.8 if you have any problems, and, by the way, 0! = 1! = 1).

Problem 1:   Write a program that will accept as input any nonnegative integer n and output n!. (Note: Since $70! > 10^{100}$, your calculator will only work up to an n of 69.)

Returning to Alice, Bill, Charlie, and Dan, how many three-person pictures can be taken? Again, the answer is 24 pictures (take all the four-person pictures and remove the person sitting on the right), the same total as the number of three-letter "words" except that different combinations of letters (made from the letters ABCD) are used.

If $1 \leq k \leq n$, the number of k-letter words (consisting of k different letters) that can be made from an alphabet of n letters is denoted by the symbol, P(n,k), and is called the number of "permutations of n things taken k at a time." It therefore symbolizes the number of ways in which k people out of n can line up for a photograph. Here is a formula for P(n,k):

$$P(n,k) = n(n - 1)(n - 2) \ldots (n - k + 1)$$

That is, P(n,k) is the product of all numbers from $(n - k + 1)$ to n, inclusive, or a product of k consecutive numbers, the largest of which is n. [If $k = n$, note that $P(n,n) = n!$, as it should.]

Problem 2: Write a program that will accept as input whole numbers k and n such that $0 \leq k \leq n$ and then output P(n,k). [This is another application of loops (see Sec. 1.8), and we leave the solution to you. Note that $P(n,0) = 1.$]

Alice, Bill, Charlie, and Dan are to choose from among themselves a committee of two to go out for beer. In how many ways can they do this? A count shows that there are six possible committees: AB, AC, AD, BC, BD, and CD. Notice that this is not the same as the number of two-letter "words" [P(4,2) = 12]. Each committee represents *two* two-letter "words;" for example, AB and BA constitute the same committee but are different two-letter "words."

When $0 \leq k \leq n$, the total number of committees of k people that can be selected from n people is denoted by either of the symbols C(n,k) or $\binom{n}{k}$. It is called the "number of combinations of n things taken k at a time" or "n choose k" or the "binomial coefficient n over k." The formula for it is

$$\binom{n}{k} = \frac{n!}{k! \, (n - k)!}$$

For example:

$$\binom{4}{0} = \frac{4!}{0!4!} = 1$$

$$\binom{4}{1} = \frac{4!}{1!3!} = 4$$

$$\binom{4}{2} = \frac{4!}{2!2!} = 6$$

$$\binom{4}{3} = \frac{4!}{3!1!} = 4$$

$$\binom{4}{4} = \frac{4!}{4!0!} = 1$$

Problem 3: Write a program that will accept as input whole numbers k and n such that $0 \leq k \leq n$ and then output $\binom{n}{k}$.

Solution to Problem 3: One way to do the problem is to set up a subroutine to calculate x! as in Problem 1, use it to calculate n!, k!, and $(n - k)!$, and then evaluate the formula, $\binom{n}{k} = \frac{n!}{k!(n - k)!}$. This method will work satisfactorily if n is small, but there will be a problem otherwise. For one thing, since it involves calculating n!, which is a lot larger than the answer, $\binom{n}{k}$, your calculator might overflow even though $\binom{n}{k}$ itself does not overflow the calculator. For example, $\binom{70}{2} = 2415$, a figure easily handled, whereas $70! > 10^{100}$ and so overflows the calculator. Furthermore, this method may prove inaccurate even if it does not make the calculator overflow. The error occurs when there are more digits in n! than the calculator can hold, causing it to round off and lose some digits.

What is needed is a method for calculating $\binom{n}{k}$ that does not involve numbers much larger than $\binom{n}{k}$ itself [see Sec. 1.2(9)]. Rather than perform all the multiplications for n! first, one might perform some of the divisions in the denominator of the fraction, $n!/k!(n - k)!$, along the way so as to keep the numbers small. But here another problem presents itself. If we do these divisions, the results might not be whole numbers despite the fact that the final answer $\binom{n}{k}$, is a whole number [see Sec. 1.2(9) again]. This can also introduce some inaccuracies.

Here is a way that seems to solve most of the problems. It will still fail if n is much too large, but it will accommodate many more inputs than the first method. Notice, to begin with, that

$$\binom{n}{n - k} = \frac{n!}{(n - k)![n - (n - k)]!} = \frac{n!}{(n - k)!k!} = \binom{n}{k}$$

For example, $\dbinom{100}{98} = \dbinom{100}{2} = 4950$. Consequently, we have a choice

in calculating $\dbinom{n}{k}$. We can calculate either $\dbinom{n}{k}$ or $\dbinom{n}{n-k}$, whichever is more convenient. It turns out that this method is quicker for
smaller k. Thus, the first step in the program is to replace k by n −
k if the latter is smaller. (Note that, in this case, $k \leq n - k$, which
implies that $k \leq n/2$.)

Now let us examine the formula,

$$\binom{n}{k} = \frac{n!}{k!\,(n-k)!}$$

Observe that the term $(n - k)!$ in the denominator is a factor of the
numerator n!. In fact, $n! = [n(n-1) \ldots (n-k+1)](n-k)! = P(n,k)(n-k)!$ Canceling out the common factor $(n-k)!$ from the
numerator and denominator in the formula $\dbinom{n}{k}$ gives

$$\binom{n}{k} = \frac{P(n,k)}{k!} = \frac{(n)(n-1)(n-2)\ldots(n-k+2)(n-k+1)}{(1)(2)(3)\ldots(k-1)(k)}$$

We can regroup to obtain

$$\binom{n}{k} = \left(\frac{n}{1}\right)\left(\frac{n-1}{2}\right)\left(\frac{n-2}{3}\right)\ldots\left(\frac{n-k+2}{k-1}\right)\left(\frac{n-k+1}{k}\right)$$

Here, then, is our method. We set up a loop that successively

accumulates in memory m_6 the following products: $p_0 = 1$, $p_1 = \left(\dfrac{n}{1}\right)$,

$p_2 = \left(\dfrac{n}{1}\right)\left(\dfrac{n-1}{2}\right)$, $p_3 = \left(\dfrac{n}{1}\right)\left(\dfrac{n-1}{2}\right)\left(\dfrac{n-2}{3}\right)$, and so on, until finally

$p_k = \left(\dfrac{n}{1}\right)\left(\dfrac{n-1}{2}\right)\left(\dfrac{n-2}{3}\right)\ldots\left(\dfrac{n-k+2}{k-1}\right)\left(\dfrac{n-k+1}{k}\right) = \dbinom{n}{k}$. The
flowchart in Fig. 2-14 should now be clear.

At each stage in this method, memory m_0 is both multiplied
and divided by some numbers. Thus, the contents of m_0 are not nearly
so large as they would be if all multiplications were done first. Also,
since m_0 will always contain integers, there will be no fractions to
affect the accuracy. The reason is that $p_0 = \dbinom{n}{0}$, $p_1 = \dbinom{n}{2}$, etc., and
all of these are whole numbers.

When you have written your program, you should experiment
with it to see for which values of n and k it is accurate and for which
values it overflows. [*Solution was realized on an HP-25 in 25 steps.*]

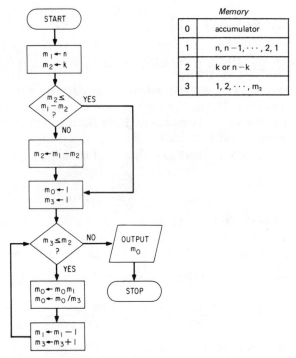

Fig. 2-14 Flowchart for outputting $\binom{n}{k}$

2.10   Multiple Precision Arithmetic

Difficulty: 3

If your calculator holds only 10 digits or so, are you forever stuck with dealing only with integers less than approximately 10^{10}? Try multiplying 2 by itself 40 times on your calculator. If the calculator is accurate to 10 places, the answer it gives is

$$2^{40} = 1.099511628 \times 10^{12} = 1,099,511,628,000.$$

Of course, this answer is not correct; in fact,

$$2^{40} = 1,099,511,627,776$$

The question is, then, can we get your calculator to deal with such large integers accurately? Indeed we can. "Multiple precision" is the technique of using several memories to store a single, large number. (Don't get this confused with "Multiple Storage," the technique of storing several small numbers in a single memory.)

Let us agree to store eight digits per memory. (Your calculator holds more than eight, but don't crowd it; we will need at least one extra digit to play with.) Allowing two registers per number, therefore, we can store numbers with as many as 16 digits. For example, we would store 2^{40} in two registers, m_0 and m_1, by having $m_1 = 10{,}995$ and $m_0 = 11{,}627{,}776$ so that $2^{40} = m_1(10^8) + m_0$.

Problem 1: Write a program to calculate the sum of two 16-digit positive whole numbers (each entered as a pair of 8-digit numbers), accurate to 17 digits.

Problem 2: Write a program that will successively calculate 2, 2^2, 2^3, ..., 2^{54}, accurate to 17 digits. (Hint: Now that we know how to add large numbers, we can easily calculate large powers of 2. Note that $2^n = 2(2^{n-1}) = 2^{n-1} + 2^{n-1}$; that is, each power of 2 can be computed by adding the previous power of 2 to itself.)

Problem 3: Here is another application. The Fibonacci sequence is the sequence of numbers beginning 1, 1, 2, 3, 5, 8, 13, ..., in which each number is the sum of the previous two. These numbers get very big very fast; the fiftieth one is already larger than 10^{10}. Write a program to calculate successively the first 79 Fibonacci numbers. Use it to verify that the seventy-ninth Fibonacci number is 14,472,334,024,676,221.

Problem 4: Repeat the first three problems, but for larger numbers. There should be room even in a small calculator to add two 24-digit numbers. If you have a powerful calculator, you should be able to handle 300-digit numbers. (Hint: A larger integer can be stored in several registers as follows: Break it up into blocks of eight digits each, starting at the decimal point, and store each block in a separate register.)

Problem 5: Write a program that multiplies two eight-digit numbers and outputs the 16-digit answer in two eight-digit blocks. (Hint: Break the numbers up into four-digit blocks.)

Solution to Problem 1: If s and t are the numbers to be added, lets begin by storing the right-most eight digits of t in m_2, and the rest of t in m_3. Thus,

$$s = m_1(10^8) + m_0$$
$$t = m_3(10^8) + m_2$$

Let x denote the (as yet unknown) right-most eight digits of (s + t); let y be the rest of (s + t) so that

$$(s + t) = y(10^8) + x$$

and x and y are the numbers we are seeking.

     Calculate the value ℓ where

$$\ell = m_0 + m_2$$

which might be a nine-digit number. The right-most eight digits of ℓ comprise the number x, and the ninth digit, c, is a "carry over." Hence, we need to separate ℓ into two parts: $\ell = c(10^8) + x$. Now if we divide ℓ by 10^8, we push the "x part" of ℓ to the right of the decimal point, leaving the digit c just left of the decimal point. Therefore, c = INT($\ell/10^8$) and x = $\ell - c(10^8)$, and we have the right-most eight digits of the answer. The rest are calculated by the equation: $y = m_1 + m_3 + c$. The flowchart is shown in Fig. 2-15.

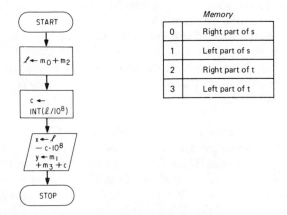

Fig. 2-15 Flowchart for calculating the sum of two 16-digit positive whole numbers accurate to 17 digits

     (Notice that this procedure is analogous to adding two two-digit numbers. We first add the right digits, writing down the right digit of the sum. The left digit is "carried over" to the next column containing the left digits. The left digit of the answer is the sum of the left digits of the summands plus the carry. In fact, what is happening above can be thought of as adding two two-digit numbers in base 10^8, the digits being the numbers m_0, m_1, m_2, and m_3. It is worth noting also that the number c is always either 0 or 1. You can verify for yourself that when you add two numbers together you always carry a 1 or you carry nothing no matter what base you are working in.)

Solution to Problem 2:   We will use registers m_1 and m_0 to store the current power of 2, as in Problem 1:

$$2^k = 10^8 m_1 + m_0$$

Since the initial power of 2, that is, 2^0, equals 1, we begin with $m_1 = 0$, $m_0 = 1$ and, set up an "endless" loop as follows: The loop begins by outputting the current power of 2, that is, the contents of m_1 and m_0. Then the next power of 2 is calculated. Since each power of 2 is just the previous one added to itself, we use the method of Problem 1 to add the current power, in m_1 and m_0, to itself and place the result back in m_1 and m_0. Then we return to the beginning of the loop. The flowchart is shown in Fig. 2-16.

Solution to Problem 3:   Since the calculator needs to remember the last two Fibonacci numbers in order to calculate the current one, we will store the most recently calculated Fibonacci number in m_2 and m_3

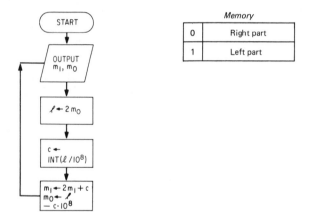

0	Right part
1	Left part

Memory

Fig. 2-16   Flowchart for successively calculating 2, 2^2, 2^3, . . . 2^{54}, accurate to 17 digits

and the one before that in m_0 and m_1. Since the first two numbers are each 1, we initialize with the values: $m_1 = 0$, $m_0 = 1$, $m_3 = 0$, and $m_2 = 1$. Now we start an "endless" loop, which begins by outputting the contents of m_1 and m_0. The calculation then goes as follows: Since the current Fibonacci number, now in m_2 and m_3, will soon become the next-to-last one, it should be moved to m_0 and m_1. An easy way to do this is first to switch the contents of m_0 and m_1 with the contents of m_2 and m_3, respectively. Since the next Fibonacci number is the sum of the two previous ones, we will use the method of Problem 1 to add the two previous numbers (still in m_0, m_1, m_2, and m_3) and place the sum in m_2 and m_3. Then we return to the top of the loop. The flowchart is shown in Fig. 2-17.

Solution to Problem 4:   To add together two large numbers, suppose the right-most block of eight digits of the first number is stored in m_0, the next block of eight in m_1, and so on. The second number is stored similarly in registers n_0, n_1, etc.

To save on registers, we won't use another whole set to store the sum; it turns out that we can store the sum back in m_0, m_1, etc. Calculate the value of $m_0 + n_0$ and split it into two pieces; the right-most eight digits, which you store back in m_0, and the ninth digit (the carry over), which you add to the contents of m_1. Repeat with $m_1 + n_1$, $m_2 + n_2$, and so on, until you get to the last pair, $m_k + n_k$, which does not need to be split; just store $m_k + n_k$ in m_k. The sum now appears in m_0, m_1, ..., m_k.

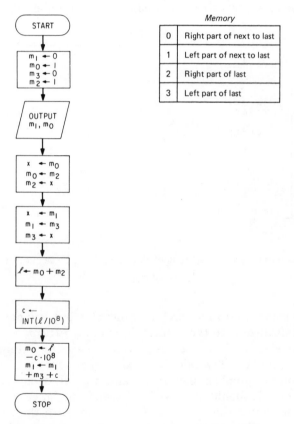

	Memory
0	Right part of next to last
1	Left part of next to last
2	Right part of last
3	Left part of last

Fig. 2-17 Flowchart for calculating successively the first 79 Fibonacci numbers

If your calculator allows subroutines and indirect addressing, they can be used to great advantage in this program, but you don't need them. We were able to add two 32-digit numbers (four registers each) on the HP-25's eight registers and 49 program steps with no trouble. The flowcharts are shown in Figs. 2-18(a) and (b).

Solution to Problem 5; Let s and t denote the two eight-digit numbers to be multiplied. We cannot use our × key to multiply them because the

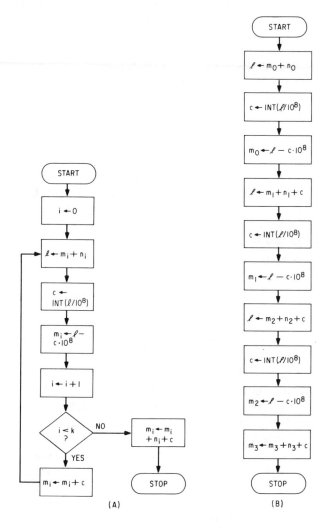

Fig. 2-18 Flowchart for Problems 1, 2, and 3 using larger numbers

calculator cannot hold all sixteen digits of the product. We begin by chopping s and t into four-digit chunks each, as follows:

$$s = s_1(10^4) + s_0$$
$$t = t_1(10^4) + t_0$$

Here, s_1 represents the four left-most digits of s, and s_0, the four right-most digits of s, and similar values hold for t_1 and t_0.

It is easy to find s_1. Dividing s by 10^4 places the decimal point in the middle, with four digits on each side. Hence the INT key, when applied to the result, produces s_1, as follows:

$$s_1 = INT(s/10^4)$$

Now, of course, $s_0 = s - s_1(10^4)$. Then, t_1 and t_0 are found similarly:

$$t_1 = INT(t/10^4)$$
$$t_0 = t - t_1(10^4)$$

To see what to do next, we determine the product, st:

$$st = [s_1(10^4) + s_0] [t_1(10^4) + t_0] = s_1 t_1(10^8) + (s_1 t_0 + s_0 t_1)(10^4) + s_0 t_0$$

Therefore, what we need are not the four numbers, s_1, s_0, t_1, and t_0, but the four numbers, $s_1 t_1$, $s_1 t_0$, $s_0 t_1$, and $s_0 t_0$. These can be easily computed using memory multiplication (in m_3, m_2, m_1, and m_0, respectively) while s_1, s_0, t_1, and t_0 are being calculated. The flowchart is shown in Fig. 2-19.

The sixteen-digit product st that we want will be calculated in two eight-digit chunks, z_1 and z_0, as follows:

$$st = z_1(10^8) + z_0$$

From our formula,

$$st = s_1 t_1(10^8) + (s_1 t_0 + s_0 t_1)(10^4) + s_0 t_0$$
$$= m_3(10^8) + (m_2 + m_1)(10^4) + m_0$$

we see that the eight-digit number m_3 is part of z_1 and the eight-digit number m_0 is part of z_0, but the eight- or nine-digit number $(m_2 + m_1)$ is straddling the fence; the right-most four digits are part of z_0 and the left-most four or five digits are part of z_1.

The next step in our program calculates ℓ, where $\ell = m_2 + m_1$, and breaks it into two chunks, ℓ_1 and ℓ_0, as follows:

$$\ell = \ell_1(10^4) + \ell_0$$

The procedure is as before:

$$\ell_1 = INT(\ell/10^4)$$
$$\ell_0 = \ell - \ell_1(10^4)$$

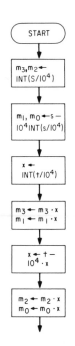

Memory	
0	$s_0 t_0$
1	$s_0 t_1$
2	$s_1 t_0$
3	$s_1 t_1$

Fig. 2-19 Flowchart for multiplying two eight-digit numbers, with the output in two eight-digit blocks (Part 1)

Returning to our main formula, we have

$$st = m_3(10^8) + \ell(10^4) + m_0$$
$$= m_3(10^8) + [\ell_1(10^4) + \ell_0] (10^4) + m_0$$
$$= (m_3 + \ell_1) (10^8) + [\ell_0(10^4) + m_0]$$

Using memory addition, the numbers $(m_3 + \ell_1)$ and $[\ell_0(10^4) + m_0]$ can be calculatd and stored in m_3 and m_0, respectively, while ℓ_1 and ℓ_0 are being calculated. The flowchart, which starts where the flowchart in Fig. 2-18 left off, is shown in Fig. 2-20.

With these new numbers in m_3 and m_0, we now have

$$st = m_3(10^8) + m_0 = z_1(10^8) + z_0$$

and thus we are almost done. We would completely be done if m_0 had eight digits at most; then, the numbers z_1 and z_0 that we are trying to calculate would be equal to m_3 and m_0, respectively. The only problem is that m_0 might be a nine-digit number. Consequently, we have to separate m_0 into two chunks, the right-most eight digits being equal to z_0 and the ninth digit, c, being carried over into m_3:

$$m_0 = c(10^8) + z_0$$

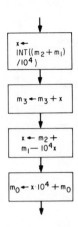

Memory	
0	$s_0 t_0 + 10^4 \, l_0$
1	$s_0 t_1$
2	$s_1 t_0$
3	$s_1 t_1 + l_1$

Fig. 2-20 Flowchart for multiplying two eight-digit numbers, with the output in two eight-digit blocks (Part 2)

As before,

$$c = INT(m_0/10^8)$$
$$z_0 = m_0 - c(10^8)$$

Returning to the main formula, we have

$$st = m_3(10^8) + m_0 = (m_3 + c)(10^8) + z_0$$

Again we use memory arithmetic to add c to m_3 and to place z_0 in m_0 while we calculate c and z_0 (see the flowchart in Fig. 2-21).

For the complete program, just string the flowcharts in Figs. 2-19, 2-20, and 2-21 together.

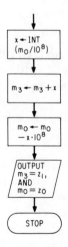

Memory	
0	z_0
3	z_1

Fig. 2-21 Flowchart for multiplying two eight-digit numbers, with the output in two eight-digit blocks (Part 3)

2.11   Infinite Precision Arithmetic

Difficulty: 3

Suppose that you want to calculate 5419351/1725033 correct to 14 digits.[6] Since your calculator probably does not hold 14 digits, some kind of trickery is called for. In the previous section we discussed a technique for doubling (approximately) the number of digits your calculator can handle. You might be able to treat this problem using double-precision arithmetic (although it wouldn't be easy), but suppose that you wanted the answer to be correct to 100 digits.

As a matter of fact, you were taught in grade school a technique for calculating fractions that will give you as many digits as you want. It's called long division, and we can use it here. Let's analyze this method. You will find that it is not too hard to teach to your calculator, and it can be set up to return digits one at a time for as long as you are willing to press the $\boxed{\text{R/S}}$ button. We will begin by discussing only proper fractions, those fractions whose denominators are larger than their numerators. When this problem has been solved, the rest will be easy.

If you think you understand long division well enough, you should not read the next paragraph and try instead to produce the program on your own. It should take as input two positive whole numbers p and q such that p < q and should return, one digit at a time, the decimal expansion of the fraction p/q. Each time the $\boxed{\text{R/S}}$ (or whatever) button is pushed, the program should return the next digit. After you have conquered this problem, you may want to tackle the following, more difficult problem.

It is well known that fractions formed from whole numbers have decimal expansions that either terminate in all zeros or repeat. For example, 3/28 = .1071428571428571428571428571428571428.... The repeating block here is 714285. Once you have the starting digits, .10, and the block, 714285, you know *all* of the infinitely many digits in the decimal expansion of 3/28. The problem is thus to produce a program that will find the starting digits and the repeating block for any fraction p/q. Note that some fractions lack one or the other of these components. For example, 1/3 starts repeating immediately (.333 ...) and 1/4 does not repeat at all (unless you want to think of it as .250000 ...). We will only sketch the solution to this more difficult problem.

Now let's take a look at long division. We will explain the process by looking at the decimal expansion of the fraction 5/28. Get

---

[6] You might wonder why you would want to do such a thing, since the numerator and denominator are only seven digit numbers. But as a matter of fact, this fraction gives the value of π correct to 14 digits.

out a pencil and paper and do the computation as it is described so that you will see what is going on. Since this is a proper fraction, the quotient begins with a decimal point. First adjoin a zero onto 5, making it 50. Next divide 28 into 50; it goes once with a remainder of 22. Thus the first digit of the decimal expansion of 5/28 is 1. To get the next digit, adjoin a zero to the remainder 22, making it 220. Now divide 28 into 220; it goes seven times with a remainder of 24, and thus 7 is the next digit in the decimal expansion. The following digit will be found by dividing 28 into 240. At each stage, the process goes like this: Adjoin a zero to the previous remainder (that is, multiply it by 10) and divide the numerator into it, getting the answer in quotient and remainder form. The heart of the technique is to determine the largest whole number of times that the denominator "goes into" the previous remainder times 10. This quotient will then be the next digit in the decimal expansion. You must also find the remainder because it will be used to generate the following digit. You might begin your solution by solving this central problem: Given two numbers a and b, find a/b in quotient and remainder form.

Now, how do you recognize when a decimal expansion is starting to repeat itself? It will start happening as soon as a remainder repeats itself. Continuing with the same example, the first two digits of 5/28 are .17. The remainder generated by the 7 was 24. Multiply this by 10 and divide by 28 to get 8 (with a remainder of 16). Continue the division (better start at the top of your paper), and you will get the following digits: 57142. The remainder generated by the 2 is 24, the same remainder as above. What will happen next? The same thing that happened the first time. A digit of 8 will be produced *and* the same remainder, 16. This will in turn generate the same 5 that was produced above, and so on. Thus the repeating block is 857142, and the beginning block is .17.

This example shows why decimal expansions of fractions (involving whole numbers) eventually repeat. Because there are only finitely many possible remainders (each remainder must be smaller than the divisor), a remainder must eventually repeat itself and force the decimal expansion to start repeating. Some decimal expansions terminate instead of repeating indefinitely. This behavior is also determined by the remainders; it happens if a remainder of zero ever occurs.

You now have enough food for thought. It is time for you to go to work.

Solution to Infinite Precision Arithmetic

First, let's consider the problem of quotient and remainder division. Given two whole numbers m and n, how do you determine the largest whole number of times that n "goes into" m and the remainder that is "left over" after finding the quotient? It's easy. First store m and n (for future use in finding the remainder). Next divide m by n and take the integer part of the result. (Think about it, and you will see that this is the whole number quotient you are looking for.) Call this number x. To find the remainder, just multiply n by x and subtract the result from m. (Think about it again. Do some examples on your calculator if you are having trouble seeing these relationships.

Now let's look at the main problem. Given p and q such that p < q, you want to turn out the digits of p/q. To do so is fairly easy, once the problem of quotient and remainder division has been solved. As explained in the statement of the problem, each digit is generated by taking the remainder from the division that produced the previous digit (to generate the *first* digit, the number p is treated as the previous remainder), multiplying it by 10, and dividing the result by q, and getting the answer in quotient and remainder form. The whole number quotient is the desired digit, and the remainder will be used to get the next digit.

Here are the details: Two memories will be used: m_1 to store q, and m_2 to store the various remainders times 10. Start with q in m_1 and p in the display. Multiply by 10, and store the contents of the display in m_2. Divide this number (which is still in the display) by q (the contents of m_1), and take the integer part of the result. This is the desired digit. Stop to display it (or you may want to just pause). Next, multiply the contents of the display by q (m_1), and subtract the result from the contents of m_2. This is the remainder used to generate the next digit; with it in the display, loop back to the step in which you multiply by 10.

The program operates as follows: The user stores q in m_1, puts p in the display, and pushes the R/S button. The program runs until it hits the stop order. It stops with the first digit of p/q in the display. The user presses R/S again, and the program generates the remainder, loops back to the top, produces the next digit, and stops again. The user presses R/S again, and so on. (See the flowchart to visualize the process.)

Now what should the program do to find p/q if p is not smaller than q? You still want to divide p by q and get the answer in quotient and remainder form. Suppose that p/q gives a (whole number) quotient of x and remainder r. Then, p/q = x + (r/q), *and* r < q. The number x will contain all digits to the left of the decimal point, and r/q will contain all digits to the right. You already have a program to generate the digits of r/q. To get x, all you have to do is start performing a quotient and remainder division. Since the program we have already performs a quotient and remainder division, it can be used to generate x too. However, you need to skip the first part of the program in which the numerator of the fraction gets multiplied by 10.

Thus the expanded program will look just like the program described three paragraphs back except for the following: It will start with a test to see if p < q; if so, it will move to the top of that program. If p ⩾ q, however, the program will branch and enter the program at the point at which the contents of the display are stored in m_2. What will happen is that on the first push of $\boxed{\text{R/S}}$ the program will stop with all digits of p/q to the left of the decimal point in the display. After that, it will generate the digits of r/q one at a time.[7] [*Solution was realized on a TI-57 in 21 steps.*]

Finally, let's have a look at the problem of finding the beginning block of digits of the fraction p/q and the repeating block. We will once again restrict our attention to proper fractions, leaving it to you to modify the program to handle improper fractions. Recall that for the fraction 5/28 the starting block of digits was 17 and the repeating block was 857142. The decimal expansion of 5/28, therefore, is .17857142857142857142. . . . Recall further that the remainders produced by the divisions that generate the digits are what determine when and where a block starts to repeat. As soon as a remainder repeats itself, we know that all digits back to the one produced by the first occurrence of the repeated remainder form the repeating block. In the case of the fraction 5/28, the first remainder is the numerator of the fraction, 5 (perhaps we should call it the zeroth remainder), which produces the first digit, 1, and in turn the second remainder, 22. The successive remainders (starting from 5) are 5, 22, 24, 16, 20, 4, 12, 8, 24, ... The first two remainders, 5 and 22, generate the starting block, .17, and the next six form a repeating block of remainders that will generate the repeating block of digits. We can diagram the succession of remainders and the corresponding succession of digits as shown in Fig. 2-22.

These diagrams should look familiar to those of you who read Sec. 1.16. They are ρ diagrams, and the algorithm described in that section can be used to find the tail and the cycle in the left-hand diagram, which in turn will produce the right-hand diagram. That diagram gives a complete description of the decimal expansion of 5/28. If you discovered this solution to the problem, give yourself an A!

There is, however, a much more efficient solution that depends on the specialized number theoretic structure of the long division algorithm. One can determine at the outset which remainder will be the first to repeat itself. In the example above, it is the third: 24. Knowing this, it is not difficult to find the starting and repeating blocks. The

---

[7] See the Notes for an alternative modification.

Fig. 2-22 Flowchart for the decimal expansion of 5/28: (a) successive remainders and (b) successive digits

starting block will consist of the two digits, 17, produced by the first two remainders, 5 and 22. To find the repeating block, store the remainder 24, which you know will eventually repeat. Then simply start turning out digits and remainders. Each time a remainder is produced, check to see if it is equal to the stored remainder. If so, you have completed the repeating block.

Now, how do you find this mysterious repeating remainder? Let p and q be the numerator and denominator of the fraction. First cancel any 2's or 5's common to both the numerator and denominator. For example, suppose that you start with the fraction 25/140. Since there is a factor of 5 in both numerator and denominator, you cancel it out, leaving the fraction 5/28. (This fraction happens to be in lowest terms already; that is, there are *no* common factors in the numerator and denominator. But it is not always necessary to get the fraction in lowest terms; you need only get rid on the common 2's and 5's. Because of the structure of the decimal number system, it is easy to recognize when these particular common factors are occurring.) Next, look at the *denominator* and count (1) how many times it is divisible by 2, and (2) how many times it is divisible by 5. Let n be the *larger* of these two numbers. In the case of our example, since 28 is divisible twice by 2 and indivisible by 5, n would be 2. It turns out that there will be n digits in the starting block of the decimal expansion of p/q, and, consequently, that the $(n + 1)^{st}$ remainder will be the first repeating remainder.

We are not in a position to explain to you why this relationship is true. It has to do with the fact that 10, the base of the decimal number system, is equal to 2 times 5. However, if you are willing to accept this relationship, we can now describe a fairly simple elaboration of the fundamental program for cranking out the digits that will determine the starting and repeating blocks.

The new program will contain the basic program for generating the digits one at a time as a subroutine (see the flowcharts in Fig. 2-23). Instead of recycling as indicated in this figure, however, the subroutine will create just one digit each time it is called and quit at

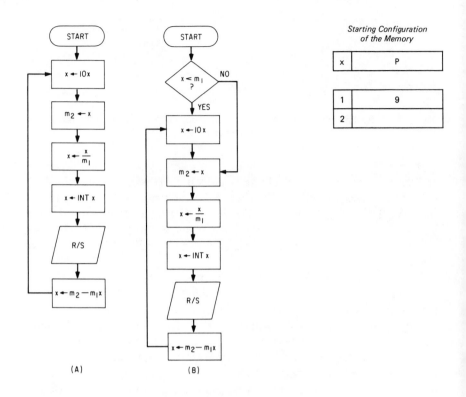

Fig. 2-23 Flowchart for generating the digits of p/q (a) with $p < q$ and (b) for any p and q that are whole numbers

the bottom, with the remainder in the display. (If your calculator does not have subroutines built in, see Sec. 1.7). The program will have two sections, A and B. Section A will generate the starting block of digits, ending with some appropriate signal—perhaps a negative number (we chose -0). Section B will generate the repeating block.

Here's how the program will work. The user will figure out the number n described in the last paragraph and store it in some memory (if you have a larger calculator, the computation of n could be automated and incorporated into the program). The program will begin with a test to see if $n = 0$. If it is, there is no starting block, and the program branches to section B. If not, go to section A, a loop for generating the starting block. Each time through the loop the digit-generating subroutine will be called once, and the contents of the memory that started with n in it will be reduced by 1. Next, run a check to see if it is down to zero. If not, go back to the top of the loop. If so, give the end-of-block signal and branch to section B. Be sure to retain the remainder from the last division. It will probably have to be stored somewhere during the transition from A to B, and perhaps also at other times. Take the remainder and store it in a special place for future reference (for example, the t-register if you have a TI calculator).

Now enter section B's loop. Each time through this loop the digit-generating subroutine will be calld once, and the resulting remainder will be checked against the stored remainder. If the two are not equal, go back to the top of the loop. If they are equal, give the end-of-block signal. That's it! We will leave it to you to figure out the details (see the flowchart in Fig. 2-24). You may want to change the R/S in the digit subroutine to a PAUSE to speed things up. [*Solution was realized on a TI-57 in 40 steps.*]

Notes: An alternate program for generating the digits of p/q where p may or may not be greater than q can be obtained by taking the first flowchart (which assumes that p < q) and simply moving the step, x

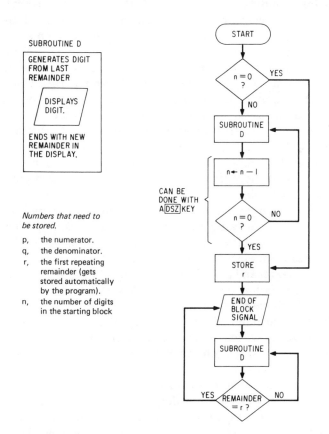

Fig. 2-24 Elaboration of the program flowcharted in Fig. 2-23

← 10x, to the bottom. This program will always turn out the digits to the left of the decimal point at the first R/S (we have done away with the test, $x < m_1$). If no such digits exist, the program will output a 0.

The processes of canceling and counting factors of 2 and 5 used in producing the number n can be facilitated by first canceling or counting the factors of 10. These are represented by zeros at the end of the numbers and therefore are easily canceled. Each factor of 10 contains one factor of 2 and one factor of 5. After the 10's have been factored out, there will be only 2's or only 5's (or neither) left. (If there were both, there would be a factor of 10.)

How large can the numbers p and q get before the accuracy of these programs breaks down? Since the programs are written for infinite precision, it wouldn't make much sense to run them on numbers that cannot be precisely specified (that is, which have more digits than your calculator will accept). In practice, this means that if, for example, your calculator will take 10 digits, you don't want to use the procedures for numbers larger than 10^{10} (the first number with 11 digits). Actually, since roundoff errors can corrupt the last digit, you should stick with numbers smaller than 2×10^9 to be completely safe.

All fractions dealt with in this problem have been assumed to have whole-number numerators and denominators. This is not as stringent an assumption as it might sound. If p and q are any numbers whatever with finitely many digits (this description certainly includes any numbers a calculator or computer can hold), then the fraction p/q can be rewritten as a fraction with a whole-number numerator and denominator. (The result, however, may be a numerator or denominator too large for these programs.) The transformation does not change the value of the fraction, and it converts both numerator and denominator into whole numbers.

The Pythagoreans (circa 500 B.C.) believed that *any* number should be expressible as a quotient of whole numbers. Fate dealt the Pythagoreans a cruel blow in the form of the number $\sqrt{2}$. Euclid's *Elements* contains an elegant proof that $\sqrt{2}$ cannot be equal to p/q for any whole numbers p and q. The proof goes as follows: We may assume that the fraction p/q has been reduced to lowest terms. Thus either p or q must be odd (if both were even, the fraction would not be in lowest terms). On the other hand, if $p/q = \sqrt{2}$, then $p^2/q^2 = 2$, or $p^2 = 2q^2$. We leave it to you to show that this relationship holds true only if both p and q are even. Thus either p or q must be *both* even and odd—a fairly decisive contradiction.

The existence of numbers not expressible as ratios of whole numbers, *irrationals* as they are called, can also be demonstrated from remarks made earlier. As we noted in the statement of the problem, if p and q are whole numbers, then the decimal expansion of p/q must eventually repeat (or terminate). Thus, if x is a number with a non-repeating decimal expansion, it cannot be a quotient of whole numbers. An example is when x = .101001000100001. . . .

2.12   Complex Arithmetic

Difficulty: 1

Complex numbers are numbers z of the form x + yi where x and y are real numbers, and i represents the square root of -1. Two sample complex numbers are 4 + 5i and $\sqrt{3}$ − i. x is called the *real part* of z, and yi (or sometimes just y) is called the *imaginary part*. The latter name is unfortunate because it suggests that there is something ficticious about the number i. In fact, complex numbers are very useful in modeling real word pheonomena, and the mathematical subject of complex variables, which means calculus involving complex numbers instead of real numbers, is one of the most fundamental branches of higher mathematics.

The object of this exercise is to get your calculator to perform ordinary arithmetic—addition, subtraction, multiplication, division, root extraction, and powers—on complex numbers. The applications manual for your calculator probably has programs for these operations, but you will learn something about complex arithmetic and about your calculator's button for converting between rectangular and polar coordinates if you work through this problem.

First we need to look at two schemes for representing complex numbers. You have already seen one, x + yi, the standard or "rectangular" form (the reason for this name will become clear shortly). To store x + yi on the calculator, two memories are needed, one for x and the other for y.[8]

Before looking at the other representation scheme, let's do a little complex arithmetic in rectangular form. Let z_1 and z_2 be two complex numbers, where $z_1 = x_1 + y_1i$ and $z_2 = x_2 + y_2i$. To find $z_1 + z_2$, $z_1 - z_2$, and z_1z_2, just pretend that you know what you're doing and carry out the computation. For example, $z_1 + z_2 = x_1 + y_1i + x_2 + y_2i = x_1 + x_2 + y_1i + y_2i = (x_1 + x_2) + (y_1 + y_2)i$. The real part of $z_1 + z_2$, then, is simply the sum of the real parts of z_1 and z_2, and likewise with the imaginary part. (If you are unfamiliar with complex numbers, it would probably be helpful for you to work out a concrete example so that you can see what is going on here.) A similar computation will

---

[8] Which memories you use for this purpose is a matter of convenience. For some manipulations, it may be convenient to use the stack (if you have an RPN calculator) or the x and t registers (if you have an algebraic calculator). These refinements we leave to you. For the present, we will assume the use of ordinary addressable memories to store all numbers.

show that $z_1 - z_2 = (x_1 - x_2) + (y_1 - y_2)i$. These two rules can be summarized as follows: To add or subtract two complex numbers, add or subtract the real or imaginary parts separately. Note that the goal in both computations is to find the real and imaginary parts of the answer. Once these are known, the answer is determined. These rules are so simple that it is not worth the trouble to write programs for them; you can carry out the computations "by hand" on the calculator.

Multiplication is a little more complicated:

$$z_1z_2 = (x_1 + y_1i)(x_2 + y_2i) = x_1x_2 + x_1y_2i + x_2y_1i + y_1y_2i^2$$
$$= (x_1x_2 - y_1y_2) + (x_1y_2 + x_2y_1)i$$

(Remember that $i^2 = -1$!). Here a program begins to be desirable. The other operations, which are even more complicated, can be facilitated by the introduction of a second notation scheme.

Just as real numbers are thought of as points on a line, complex numbers, which require two real numbers for their description, are thought of as points in a plane. The complex number $x + yi$ is identified with the point (x,y), x and y being called the *rectangular coordinates* of the point (x,y). Now, the same point can be described via two other numbers: ρ, the distance from the point to the origin, and θ, the angle that the line from the point to the origin makes with the positive x-axis (see Fig. 2-25). These numbers, ρ and θ, are called

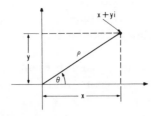

Fig. 2-25   Rectangular and polar coordinates for the complex number $x + yi$

the *polar coordinates* of the point. Not surprisingly, ρ and θ can also be used to describe the complex number $x + yi$. To see exactly how this is done, we need to look at the conversion formulas between rectangular and polar coordinates.

By examining Fig. 2-25 and recalling a little trigonometry, you will see that $x = \rho\cos\theta$, and $y = \rho\sin\theta$. Thus, the complex number $z = x + yi$ can be rewritten in terms of ρ and θ as $\rho\cos\theta + i\rho\sin\theta$, or $\rho(\cos\theta + i\sin\theta)$. This is called the *polar form* of the complex number z, ρ being called the *absolute value* of z and θ the *argument*.

To see the significance of polar form for complex arithmetic, one more formula is needed. It is DeMoivre's formula[9]:

$$e^{i\theta} = \cos\theta + i\sin\theta$$

In the expression on the left-hand side of the equation, e represents the base of the natural logarithms. From DeMoivre's formula, you can see that the complex number, $z = \rho(\cos\theta + i\sin\theta)$, can be rewritten as follows: $z = \rho e^{i\theta}$ (this form is also referred to as *polar form*). Now comes the good part. Suppose that you have two complex numbers in the above form, $z = \rho_1 e^{i\theta_1}$ and $z_2 = \rho_2 e^{i\theta_2}$, and you want to multiply them together. It's easy to do so: $z_1 z_2 = \rho_1 e^{i\theta_1} \rho_2 e^{i\theta_2} = \rho_1 \rho_2 e^{i\theta_1 + i\theta_2} = \rho_1 \rho_2 e^{i(\theta_1 + \theta_2)}$. To multiply two complex numbers in polar form, then, all you have do is multiply absolute values and add arguments. Division works similarly. To divide two complex numbers in polar form, divide absolute values and subtract arguments.

Now this is all great *if* the numbers you are dealing with are in polar form, but suppose that they are not. Here is where your calculator's $\boxed{\text{R} \rightarrow \text{P}}$ button (or $\boxed{\text{INV}}$ $\boxed{\text{P} \rightarrow \text{R}}$ on TI calculators) comes in. This button takes the two numbers x and y and returns the two numbers ρ and θ. Another key, $\boxed{\text{P} \rightarrow \text{R}}$, goes the other way: Given ρ and θ, it returns x and y. These operations have several important uses. The one we are presently interested in is that they allow you to convert back and forth between rectangular and polar form. Thus, you can perform the arithmetic in whichever form is more convenient, converting to the other form if necessary (for example, if the problem is stated in the other form).

One example will show what we mean; then it will be up to you to see the general strategy and produce the necessary programs. Suppose that you want to multiply the numbers, $z_1 = 2 + 3i$ and $z_2 = 5 - 2i$, taking advantage of the simpler technique for multiplication in polar form. Convert the rectangular coordinates $(2,3)$ and $(5,-2)$ into polar coordinates (get out your calculator), with the following results: $\rho_1 = 3.605551275$, $\theta_1 = 56.30993247°$, $\rho_2 = 5.385164807$, and $\theta_2 = -21.80140949°$. Thus, the absolute value of $2 + 3i$ is 3.605551275 and the argument is 56.30993247°; similarly for $5 - 2i$ (you do not have to write any of these numbers down, of course; use your calculator's memories). To get the absolute value ρ and argument θ of the product $z_1 z_2$, proceed as indicated two paragraphs back—multiply absolute values and add arguments: $\rho = \rho_1 \rho_2 = 19.41648784$ and $\theta = \theta_1 + \theta_2 = 34.50852299$. You now know the product in polar form (that

---

[9] The formulas for complex arithmetic that we are about to derive using DeMoivre's formula could be obtained without it by using some elementary trigonometric identities. If you don't know DeMoivre's formula, however, you should learn it anyway (you will thank us later).

is, you know ρ and θ). All that remains is to convert ρ and θ back to x and y (use your calculator again). You will find that x = 11 and y = 16. Thus z_1z_2 = x + yi = 11 + 16i. You can verify by hand that this answer is correct.

It is possible to do multiplication and division without conversion to polar form, of course [to find $(x_1 + y_1i)/(x_2 + y_2i)$, multiply numerator and denominator by $(x_2 - y_2i)$], but fewer program steps are involved if you do convert. Polar conversion is even more useful when raising numbers to powers and extracting roots: $z^n = (\rho e^{i\theta})^n = \rho^n(e^{i\theta})^n = \rho^n e^{in\theta}$. To raise z to the nth power, raise the absolute value to the nth power and multiply the argument by n. We leave it to you to figure out how to take n^{th} roots (which is the same as raising to the 1/n^{th} power). Now it is time for you to get into the act with all of this new-found knowledge and write programs for the operations of complex multiplication, division, powers, and roots.

Solutions to Problems

The scheme for solving all these problems is basically the same:

1. Convert the rectangular coordinates for the point or points representing the complex numbers into polar coordinates.
2. Perform the indicated operations on the polar coordinates to get the polar coordinates for the point representing the answer.
3. Convert the polar coordinates back to rectangular coordinates.

We will examine the program for multiplication in detail (see Fig. 2-25). For the other programs we will give only the flowcharts (see Figs. 2-26, 2-27, and 2-28).

Let the two numbers that you want to multiply together be the following: $z_1 = x_1 + iy_1$ and $z_2 = x_2 + iy_2$. We begin by assuming that x_1, y_1, x_2, and y_2 have been stored in memories m_1, m_2, m_3, and m_4, respectively. How you get them there is up to you (see Sec. 1.5). By now you should know how to convert from rectangular to polar coordinates on your calculator. Be sure that you know where x and y go in and where ρ and θ come out. The program should start by loading x_1 and y_1 and then converting to polar coordinates ρ_1 and θ_1. Store these numbers back in m_0 and m_1 (you won't need x_1 and y_1 again). Next, load x_2 and y_2 and convert to polor coordinates ρ_2 and θ_2. Now you want to multiply ρ_1 and ρ_2 and add θ_1 and θ_2. This is probably most easily accomplished with register arithmetic (HP: STO × 0 or STO + 1; TI: SUM 0 or Prd 1). In any case, you now have $\rho_1\rho_2$, the absolute value of z_1z_2, and the argument, $\theta_1 + \theta_2$. Load these two numbers and convert from polar back to rectangular form, thus obtaining

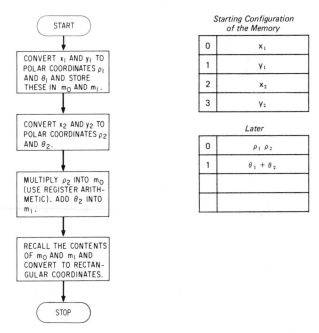

Fig. 2-26 Flowchart for the multiplication of complex numbers

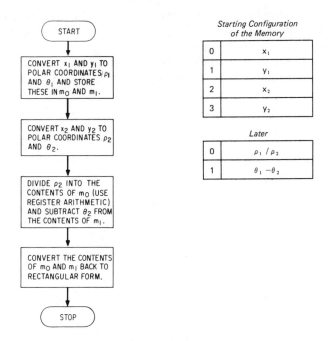

Fig. 2-27 Flowchart for the division of complex numbers

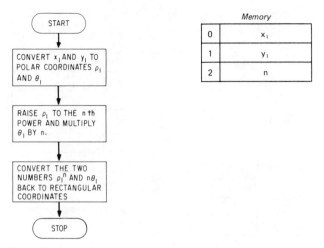

Fig. 2-28 Flowchart for raising complex numbers to a power

the real and imaginary parts of $z_1 z_2$. [*Solution was realized on a TI-58 in 33 steps.*]

Notes: The technique used in these programs of translating the problem to be solved into another language (polar form) from the one in which it was posed (rectangular form), solving the problem in the new language, and then translating back to the original language is one of the *big ideas* in mathematics.

DeMoivre's formula can be used to verify one of those equations that still make grown mathematicians think there is something

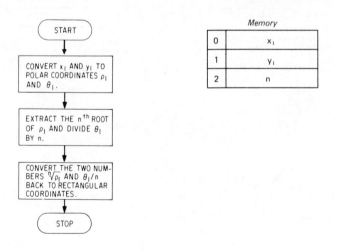

Fig. 2-29 Flowchart for extracting roots of complex numbers

magical about mathematics. This equation involves each of the five most fundamental numbers in mathematics exactly once:

$$e^{i\pi} + 1 = 0$$

(As always in calculus, π is thought of in radians).[10]

2.13 Egyptian Fractions and the Greedy Algorithm

Difficulty: 4

This problem has to do with rational numbers and the means of representing them. For instance, 3/4 can be thought of in two ways: as a *number* (the number .75 on your calculator) or as a string of symbols (3, /, 4) that *represent* that number. Since we need to distinguish between the number and its representation in what follows, we will use the word *fraction* to mean the representation: a pair of integers (whole numbers) separated by a diagonal bar and with the right-hand integer nonzero. Thus, 2/3, $-2/-3$, 34/51, and 22/7 are fractions (whereas 3/0, 3.1/1, and $\sqrt{2}/\sqrt{18}$ are not. The *value* of a fraction will mean the number obtained by dividing the right-hand integer (the *denominator*) into the left-hand (the *numerator*).

A number is said to be *rational* if it is the value of a fraction. Thus, 2/3, $-2/-3$, and 34/51 are different fractions, but they all have the same value, that is, that all represent the same rational number, 2/3. Though 3.1/1 and $\sqrt{2}/\sqrt{18}$ are not fractions, they represent rational numbers since $3.1/1 = 31/10$ and $\sqrt{2}/\sqrt{18} = \sqrt{2}/\sqrt{2}\,\sqrt{9} = 1/\sqrt{9} = 1/3$. Any integer is a rational number; for example, $10 = 10/1$, $-5 = -5/1$, and $0 = 0/1$. (But many numbers are *not* rational. For example, it is impossible to find integers a and b such that $a/b = \sqrt{2}$. Consequently, $\sqrt{2}$ is not rational (see the Note in Sec. 2.11). Other examples of irrational numbers are $\sqrt{3}$, $\sqrt{17}$, π, and e.

The ancient Egyptians used only rational numbers in their arithmetic and they represented these numbers in a very unusual way, namely, as sums of *unit fractions*, that is, fractions of the form 1/n. Moreover, they never used the same unit fraction more than once in a sum. They might have written 2/5, for example, as $(1/5) + (1/10) + (1/15) + (1/30)$. There are many ways of expressing a positive rational number as a sum of different unit fractions; for an illustration, verify that 2/5 may also be written as $(1/5) + (1/10) + (1/15) + (1/60) + (1/90) + (1/180)$ and as $(1/3) + (1/15)$.

It is a fact, though not at all an obvious one, that any positive rational number can be written as a sum of finitely many different

---

[10] Note that the equation is the identity that results when $\theta = \pi$ in DeMoivre's formula. It is credited to Euler.

unit fractions. We shall describe a method due to Leonardo of Pisa, a mathematician of the early thirteenth century (better known as Fibonacci—the same Fibonacci who formulated the Fibonacci sequence described in Sec. 2.4). We can illustrate his method with the rational 55/34. What we need is to find a finite sequence $n_1, n_2, \ldots,$ of positive integers such that $55/34 = (1/n_1) + (1/n_2) + \ldots$. Since the numbers $n_1, n_2, \ldots$ are all supposed to be different, we might as well put them in increasing order: $n_1 < n_2 < \ldots$.

Fibonacci's technique is to find the numbers $n_1, n_2, \ldots$ one by one. What must be true about n_1? Well, certainly, $1 \leq n_1$. Moreover, $1/n_1$ cannot be bigger than 55/34, that is, $1/n_1 \leq 55/34$, or $34/55 \leq n_1$. Now comes the main idea. Why not be greedy and choose n_1 so that $1/n_1$ is as big as possible? That will leave the smallest possible rational, $(55/34) - (1/n_1)$, left over to be expressed as $(1/n_2) + (1/n_3) + \ldots$. For $1/n_1$ to be as big as possible, n_1 must be as small as possible. Therefore, we choose n_1 to be the smallest integer satisfying the relationships, $1 \leq n_1$ and $34/55 \leq n_1$. Surely, the smallest possible n_1 is 1. This gives us the following: $55/34 = (1/1) + (1/n_2) + \ldots$, or $(55/34) - 1 = 21/34 = (1/n_2) + (1/n_3) + \ldots$.

Now we work on n_2. Surely $n_2 > n_1$, or $2 \leq n_2$. Moreover, $1/n_2 \leq 21/34$, or $34/21 \leq n_2$. Again, we'll be as greedy as we can. We choose n_2 to be the smallest integer satisfying the relationships, $2 \leq n_2$ and $34/21 \leq n_2$. Clearly, n_2 should equal 2. Thus, $21/34 = (1/2) + (1/n_3) + \ldots$, or $2/17 = 4/34 = (21/34) - (1/2) = (1/n_3) + (1/n_4) + \ldots$. What about n_3? Well, since we want $n_2 < n_3$, $3 \leq n_3$. Furthermore, $1/n_3 \leq 2/17$, or $17/2 \leq n_3$. The smallest possible such n_3 is 9. Thus, $2/17 = (1/9) + (1/n_4) + \ldots$, or $1/153 = (2/17) - (1/9) = (1/n_4) + (1/n_5) + \ldots$. For n_4, we want the smallest integer satisfying the relationships, $n_4 > n_3$ (or $10 \leq n_4$) and $1/n_4 \leq 1/153$ (or $153 \leq n_4$). We choose n_4 to be 153, and this gives us the following: $1/153 = (1/153) + (1/n_5) + \ldots$, or $0 = (1/153) - (1/153) = (1/n_5) + \ldots$. And we are therefore done; there is no $n_5, n_6, \ldots$. Thus, $55/34 = (1/1) + (1/2) + (1/9) + (1/153)$, as you can check.

Problem: Write a program that accepts as input a pair (p,q) of positive integers and uses Fibonacci's technique to output a sequence $n_1, n_2, \ldots$ of positive integers such that $p/q = (1/n_1) + (1/n_2) + \ldots$.

Does Fibonacci's technique always work? There is only one way it could fail—if it never stopped. Conceivably, for some rational p/q the algorithm will continue to crank out ever larger numbers $n_1, n_2, n_3, \ldots$ and never stop. (In this case, we would still have the desired expression of p/q in some sense, but the problem requires writing p/q as a sum of *finitely* many unit fractions.) In fact, this situation will never occur, for the algorithm will always stop after finitely many terms are computed. (The proof is not difficult; however, it will lead us astray to give it here. The trick is to show that the numerators of the "leftovers"—55/34, 21/34, 2/17, and 1/153 in the example—always decrease.)

Many mathematical problems can be solved by an algorithm, a well-defined, step-by-step procedure that eventually produces the answer. An algorithm often involves a decision at each stage of the solution. For example, at the ith stage of Fibonacci's algorithm, we had to decide which number to choose as n_i. In many problems, one might have to evaluate each decision in terms of how it will affect several later decisions, or even the whole algorithm. As in chess or checkers, you often have to think ahead several moves before deciding what move to make.

In some problems, however, it pays to be greedy. When a decision is called for, don't worry about thinking ahead; live for the moment and grab as much as you can. Such algorithms are called "greedy algorithms." Fibonacci's technique is an example of one. When we choose n_i at the ith stage to be as small as possible, we made $1/n_1$ as big as possible and hence gobbled up as much of the number p/q as we could.

It is surprising when greed pays because in most problems it won't. Consider the following typical problem: A truck that can take a load of 100 pounds at most pulls up to a roadside pumpkin stand where several pumpkins of differing weights are for sale. The driver wishes to buy as many pounds of pumpkin as he can get in his truck. Using the greedy algorithm, he begins by buying the heaviest pumpkin under the 100-pound limit. He continues to buy pumpkins one at a time, each time buying the heaviest pumpkin whose weight, together with the weight of the pumpkins already bought, does not exceed 100 pounds. He stops when this process is no longer possible.

Does greed pay here? It does not. Suppose that only one 50-pound pumpkin and several 20-pound pumpkins are available. The greedy algorithm chooses the 50-pound pumpkin first because it is the heaviest. Since only 20-pound pumpkins are left now, two more can be chosen at the second and third stages. Since the truck load now totals 90 pounds (50 + 20 + 20), no more pumpkins can be bought.

If the driver had not been so greedy at first and thought ahead, he would have ignored the 50-pound pumpkin and bought five 20-pound pumpkins instead, giving him the maximum load of 100 pounds.

Solution: One way of storing a rational number in your calculator is to store its value in a single memory (2/3 would be stored as .6666666. . .). It is usually more accurate, however, to store numerator and denominator in two separate registers. Initially, we use registers r_0 and r_1 for numerator p and denominator q, respectively, of the given fraction, p/q. But as the program continues, r_0 and r_1 are used to store the successive "leftovers" described above:

$$\frac{p_1}{q_1} = \frac{p}{q} - \frac{1}{n_1}$$

$$\frac{p_2}{q_2} = \frac{p_1}{q_1} - \frac{1}{n_2} = \frac{p}{q} - \frac{1}{n_1} - \frac{1}{n_2}$$

$$\frac{p_3}{q_3} = \frac{p_2}{q_2} - \frac{1}{n_3} = \frac{p}{q} - \frac{1}{n_1} - \frac{1}{n_2} - \frac{1}{n_3}$$

$$\cdot$$
$$\cdot$$
$$\cdot$$

At each stage, the number $n_i + 1$ is used to pick the next n_{i+1}; we store each successive $n_i + 1$ in r_2, beginning with $r_2 = 1$. The flowchart is shown in Fig. 2-30.

At the ith stage, the rational p_i/q_i is stored in r_0 and r_1, and $n_i + 1$ is stored in r_2. We must pick n_{i+1} to be the smallest integer satisfying the two equations,

$$\frac{r_1}{r_0} = \frac{q_i}{p_i} \leqslant n_{i+1}$$

$$r_2 = n_i + 1 \leqslant n_{i+1}$$

The smallest integer x satisfying the equation,

$$\frac{p_i}{q_i} \leqslant x$$

is

$$x = \left\lceil \frac{r_1}{r_0} \right\rceil$$

(The function $\lceil \ \rceil$, "round up," was described in Sec. 1.12. Therefore, the smallest n_{i+1} satisfying both constraints is

$$n_{i+1} = \max\left(\left\lceil \frac{r_1}{r_0} \right\rceil, r_2 \right)$$

which accounts for the first three steps following the negative response to the test, $r_0 = 0?$, in Fig. 2-29 [Max (a,b) stands for the larger of the two numbers, a and b].[11] The next three steps update registers r_0, r_1, and r_2 for the next stage. We want

---

[11] See Sec. 1.2(6) for a fast method for getting max(a,b) on some calculators.

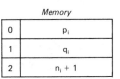

Memory	
0	p_i
1	q_i
2	$n_i + 1$

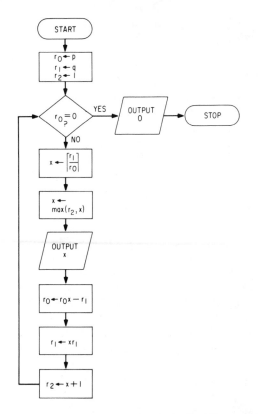

Fig. 2-30 Flowchart for the Fibonacci technique with the output:
$$p/q = (1/n_1) + (1/n_2) + \ldots$$

$$\frac{r_0}{r_1} = \frac{p_{i+1}}{q_{i+1}} = \frac{p_i}{q_i} - \frac{1}{n_{i+1}} = \frac{n_{i+1} \; p_i - q_i}{q_i \; n_{i+1}}$$

or

$$r_0 = p_{i+1} = n_{i+1}p_i - q_i$$

and

$$r_1 = q_{i+1} = n_{i+1}q_i$$

Finally,

$$r_2 = n_{i+1} + 1$$

The program now returns to the top of the loop, where the test, $r_0 = 0$? determines if we are finished or not. The program outputs a 0 to signal that it is done. [*Solution was realized on an HP-25 in 30 steps.*]

Note: No one knows why the Egyptians wanted to write all rational numbers as sums of unit fractions.

2.14   Backtrack Algorithms

Difficulty: 4   Calculator size: Medium

As we saw in Sec. 1.16, a clever trick or two can go a long way in simplifying the search for a solution to a problem. Many problems, however, are so hard that you might not be able to find any clever tricks. (Some problems are so hard, in fact, that mathematicians have proved that not even clever tricks can solve them!) For seemingly intractable problems, you are left with only one alternative: You must program your calculator to check every single possibility, carefully and tediously, until it finds a solution, or even all solutions, should you need them.

In this section, we will describe a procedure for programming such a search, one that is suitable for many kinds of problems. The particular problem we will examine is taken from the game of chess.

A queen is a chess piece that can move any number of squares horizontally, vertically, or diagonally. Our problem is this: How many ways may four queens be placed on a four-square by four-square (4×4) chess board so that no two of them are in a position to attack one another directly (that is, *in one move*). It so happens that there are 1820 ways of placing four queens on a 4×4 board. In most of them, of course, there will be queens in a direct-attack position. One obvious way to solve the problem is to have your calculator generate the 1820 ways one after the other, checking each to see if any two queens are in this position.

There is a better method, however. In the placement shown in Fig. 2-31, since the two queens already on the board are in a direct-attack position, they cannot be part of any solution. But of the 1820 ways of placing four queens on the board, 91 will place two of them in just such a way. Thus we know that it would be a waste of time for the calculator to check any of these 91 cases.

The idea, then, is to place the four queens on the board not all at once, but one at a time. If the first two queens placed are in a direct-attack position, we don't bother to place the other two but move one of the two queens already on the board. Similarly, if two of the three

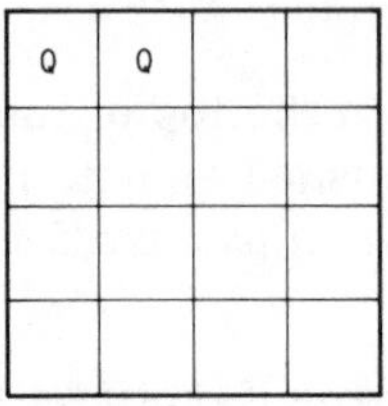

Fig. 2-31 One of the 91 placements of queens incapable of a solution

queens now on the board are in such a position, we don't place the fourth but again shift a queen already on the board. Moreover, both the trial placing of a new queen and the moving of a previously placed queen must be done in a systematic manner, for two reasons. First, we don't want the program to miss any possible solutions. Second, we don't want the program to waste time checking the same position twice.

Here is our system. Since none of the four rows on the board can contain two queens, each row must contain exactly one queen. We will thus begin by placing the first queen in the first row, the second queen in the second row, and so forth. Each time a queen is placed, it will be at the extreme left of her row. Each time she is moved, it will be one square to the right. If she falls off the right end of her row, we will "backtrack," that is, we will retreat one row up and shift the now bottommost queen.

Let us now formulate this procedure precisely. We strongly recommend that you draw a 4 × 4 board on a piece of paper, obtain four counters representing queens, and carefully comply with the directions of the following algorithm (note that for the placement of the first queen, the algorithm proceeds directly from Step 0 to Step 2):

Step 0: Start with an empty board.

Step 1: If there are four queens on the board, go to Step 6. *Otherwise,* go to Step 2.

Step 2: Place a queen in the leftmost square of the topmost empty row.

Step 3: If the queen in the bottommost occupied row does not occupy a direct-attack position against any other queen, go to Step 1. *Otherwise,* go to Step 4.

Step 4: If the queen in the bottommost occupied row is not in the rightmost square of that row, move her one square to the right, and go to Step 3. *Otherwise,* go to Step 5.

Step 5: Remove the queen in the bottommost occupied row. If there are now no queens on the board, stop; all solutions have been recorded. *Otherwise,* go to Step 4.

Step 6: Record the position of the queens; it is a solution. Go to Step 4.

Following these directions, you should record exactly two solutions, as shown in Fig. 2-32. It wouldn't hurt to run through the above procedure several times so that you will understand it better. You can't hope to teach your calculator what to do until you know how to do it yourself!

There is nothing special about the number 4 in the use of four queens on a 4 × 4 board. We can use the same procedure to find all ways of placing n queens (where n is any positive integer) on an n ×

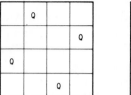

Fig. 2-32 Two solutions

n board so that no two are able to attack one another directly. Just start with n counters and an empty n × n board, and replace the 4 in Step 1 by n. (But don't try it with an n bigger than 8 unless you have lots of spare time!)

Since you will probably tire fast of moving counters around the board, it only makes sense to teach your calculator how to do it.

Problem: Write a program that accepts as input a positive integer n such that n ≤ 10 and that outputs all ways of placing n queens on an n × n board so that no two queens are able to attack one another directly, using the procedure described above.

This one is a bit of a toughie. Although the obstacles are not hard to overcome, there are a great many of them. Consequently, we will give the solution in a series of hints that are planned to overcome one obstacle at a time. If you think you know how to start off, don't read any further; try to find a solution on your own. If you get stuck, read just one more hint and try again.

Solution: *Hint 1:* Your calculator deals with numbers, but this problem is about queens on chessboards. To make your calculator capable of dealing with chessboards, you must find some way of having it store the positions of the queens.

To simplify the discussion, we shall consider only the special case for which n = 4. At any stage of the procedure, there are either no queens on the board or as many as four. Moreover, they can't be positioned just anywhere. For one thing, no two of them can ever be in the same row. Furthermore, since the rows are filled from top to bottom, it can never be the case that an empty row will be above an occupied row. (If you don't understand these observations, go back and run the procedure by hand a few more times.)

We can store the position of the queens as a number in a single register, m_2, as follows: The leftmost digit of m_2 indicates which column the queen in the top row occupies; the next digit to the right indicates which column the queen in the second row down occupies; and so forth. For example, when $m_2 = 24$, it means that there is a

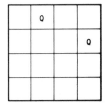

Fig. 2-33 Placement of queens when $m_2 = 24$

queen in the second column of the top row and a queen in the fourth column of the second row (see Fig. 2-33). The two solutions in Fig. 2-32 would therefore be represented by the numbers 2413 and 3142. Likewise, a single queen in the upper right corner would be represented by the single digit 4, whereas an m_2 of 0 would denote an empty board.

Hint 2: The program has to adjust positions on the board in several different ways. It must add new queens, move queens, and remove queens as required. To this end, we will use register m_2 to store the positions of the queens *currently* on the board. We call these queens the *old queens*. Since it will *never* be the case that a pair of old queens attack one another, we must make sure that any queens placed in m_2 do not attack the ones who are already there. We will also need register m_0 to store the number of old queens.

Register m_1 will consist of a single digit representing the position of a *new queen*. Since we know which row she will have to go in—the topmost unoccupied row—the digit in m_1 represents the *column* into which we are attempting to place her.

Here is how to perform four of the steps in the algorithm of the previous problem:

Step 2: To place a new queen in the leftmost square of the topmost empty row, use

$$m_1 \leftarrow 1$$

(Don't put her in with the old queens in m_2; she hasn't been tested yet.)

Step 3: If the new queen passes the test here, she must be placed in m_2 with the old queens. The operation

$$m_2 \leftarrow 10m_2 + m_1$$

moves the digits in m_2 one place to the left so as to make room for the new queen and then takes her on. The step

$$m_0 \leftarrow m_0 + 1$$

is used to tell the calculator that there is now one more old queen on the board.

Step 4:   To move the new queen one step to the right, use

$$m_1 \leftarrow m_1 + 1$$

Step 5:   This is the "backtrack" step in which the new queen having left the board, the bottommost old queen must be removed from m_2 and rejuvenated into a new queen, her column number then being recorded in m_1. The steps are as follows:

$$m_2 \leftarrow m_2/10$$
$$x \leftarrow FRAC(m_2)$$
$$m_2 \leftarrow m_2 - x$$
$$m_1 \leftarrow 10x$$

We must also tell the calculator that there is now one old queen fewer by using

$$m_0 \leftarrow m_0 - 1$$

We have given you all the steps you need to know except for the hardest one: How do we test to find out whether the new queen will attack any of the old ones? If you can't work out a solution of your own, read the next hint.

Hint 3: Let us suppose that m_2 contains the k digit number represented by $a_k a_{k-1} \ldots a_2 a_1$, which signifies that there are k old queens on the board, occupying the first k rows. The old queen in the top row is in column a_k; the old queen in the second row down is in column a_{k-1}; and so on, to the old queen in row k and column a_1. If the new queen is in column j (that is, $m_1 = j$), we must test to see whether she will attack any of the k old queens.

A diagram of the squares the new queen would attack is shown in Fig. 2-34. Notice that she will attack the bottommost old queen if

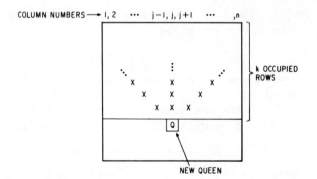

Fig. 2-34   Diagram of the squares the new queen would attack

and only if $a_1 = j - 1$, j, or $j + 1$. She would attack the next-to-last queen in column a_2 if and only if $a_2 = j - 2$, j, or $j + 2$. In general, for each i in the range $1 \leq i \leq k$, the new queen would attack the old queen in column a_i if and only if $a_i = j - i$, j, or $j + i$.

So here is our strategy. We will place the contents of m_2 into a "chopping block"—register m_3. From m_3, the digits a_1, a_2, ..., a_k are chopped off, and we then test to see if $a_i = j - i$, j, or $j + i$. If it is, then the new queen will attack the old queen in column a_i, and we must terminate this routine and go to Step 4, as instructed by Step 3 of the original algorithm. Otherwise, we continue testing by chopping off the next digit from m_3. When we have chopped off all digits (that is, when $m_3 = 0$) without finding an old queen that has been attacked by a new queen, we must return to Step 1, as instructed by Step 3 of the original algorithm. Naturally, we must use a loop to accomplish all this, as shown in the flowchart of Fig. 2-35. Note that we need another memory, m_4, to store i.

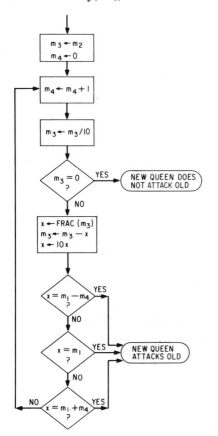

Memory	
0	# of old queens
1	New queen
2	Old queens
3	Chopping block
4	i counter

Fig. 2-35 Loop for testing the attack of the new queen

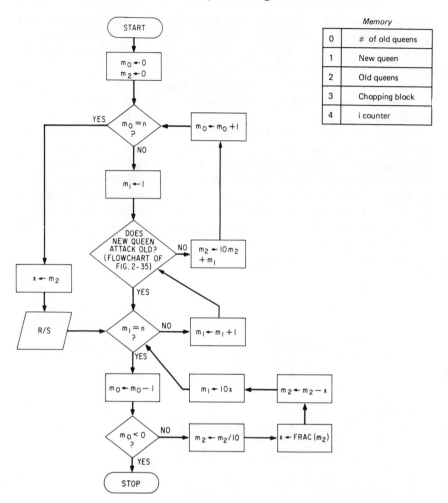

Fig. 2-36 Final flowchart for the four queens chess program

Hint 4: Now let's put all the pieces of the puzzle together. The final flowchart for the program is shown in Fig. 2-36. The flowchart of Fig. 2-37 compartmentalizes Fig. 2-36 to show which parts correspond to Steps 0 through 6 of the original algorithm.

2.15 Sorting

Difficulty: 3

To "sort" a collection of objects means to rearrange them into some prespecified order. It is a procedure frequently called for in com-

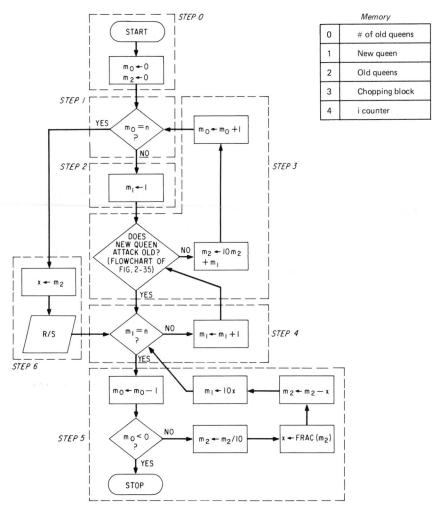

Memory	
0	# of old queens
1	New queen
2	Old queens
3	Chopping block
4	i counter

Fig. 2-37 Those parts of Fig. 2-36 that correspond to Steps 0 through 6 of the original algorithm

puting, especially when a computer is being used to store large lists of data that come to it in an unsorted fashion. Consider, for example, a business that uses a computer to keep a list of its customers. The customers' names are fed to the computer more or less randomly, but if they are to be readily retrieved, it would be a good idea to have them in alphabetical order, and the computer needs to be able to produce this rearrangement. Or consider a professor who keeps her students' test scores on a computer. For making up the grade curve, she wants to see the scores listed from lowest to highest. The scores, then, need to be sorted.

Since sorting is usually required only when large amounts of data are involved, it is not a process absolutely essential to a calculator. Nevertheless, it makes an interesting exercise.

Problem: Assume that your calculator has eight memories (if it doesn't, pretend for the moment that it does). Write a program that rearranges the contents of all but one of the memories in descending order. We leave you one remaining memory to use as you need in the program (if you own a TI-57, you will *have to* keep m_7 free; it is the test register, and clearly this problem is going to require comparisons). If m_7 is kept free, the program should work in the following manner. No matter what numbers are stored in m_0 through m_6 originally, the largest number after the program has been run will be in m_0, the next largest in m_1, and so on, with the smallest in m_6. Thus, if the memories looked like this before running the program:

0	15
1	34
2	25
3	6
4	0
5	−8
6	5280
7	

they should look like this after running the program:

0	5280
1	34
2	25
3	15
4	6
5	0
6	−8
7	

If your calculator has many memories, try this problem first and then see how far you can expand your program. If your calculator has indirect addressing, you should be able to sort the contents of all, or almost all, its memories.

Solution: The solution we are going to describe is not the most efficient as far as time of execution is concerned (see the Notes and Sec.

1.16). It is, however, relatively easy to describe and can be programmed onto a small calculator. The heart of the procedure is to begin the program by recalling the contents of m_6 and m_5, moving one into the t-register (test register), and leaving the other in the x-register (if you have a reverse Polish calculator, all you have to do is key RCL 6 and then RCL 5).[12] Next, move the larger of these two numbers into the t-register (or leave it there if there already) and the smaller into the x-register. Store the contents of the x-register in m_6. Note that the smaller of the two numbers originally in m_5 and m_6 is now in m_6 and that the larger is in the t-register.

Next recall the contents of m_4 into the x-register. Compare the x- and t-registers and rearrange their contents (if necessary) so that the larger number is in the t-register and the smaller in the x-register. Store the latter number in m_5. The larger of the three numbers originally in m_4, m_5, and m_6 will now be in the t-register and the other two in m_5 and m_6 (not necessarily in descending order). Next, recall the contents of m_3 and compare them with the contents of the t-register, moving the larger of these two numbers into the t-register and storing the smaller in m_4. Continue in this fashion up through the memory registers until you have recalled the contents of m_0, compared them with that of the t-register, stored the smaller of the two numbers in m_1, and moved the larger into the t-register. The t-register will now contain the largest of all numbers originally stored in any of the memories, m_0 through m_6, and the remaining numbers will be in m_1 through m_6. The contents of the t-register should therefore be placed in m_0.

Notice that the largest number got picked up at some point and floated to the top of the list in the t-register. For this reason, the algorithm is called a "bubble sort."

We are not done yet, however, for the numbers in m_1 through m_6 are not necessarily in their correct order. If you think about it (and we advise you to do so), you will see that although it is possible for a number to move all the way to the top of the list during the bubble sort, it can move *down* no more than one step. Clearly, some numbers will have to move down further still. How can we make them do so?

Consider what will happen if we run through exactly the same routine again. No comparisons involving the largest number, now in m_0, will occur until the end. Thus, the *second* largest number will float upward in the t-register until it finally gets compared with the contents of m_0, at which point it will get stored in m_1—exactly where we want it to go. (Note also that the contents of m_0 will first go into the

---

[12] Since many comparisons are going to be made in this program, we will have frequent occasion to refer to the t-register. If you have an HP calculator, read "y-register" for "t-register."

t-register and then right back into m_0). Run the procedure again, and the third largest number will float up and get deposited in m_2 (the contents of m_0 and m_1 staying where they are). Run the procedure a total of six times, and the numbers will all wind up in their proper positions. (We leave it to you to figure out why six runs are sufficient to get the seven numbers arranged in their proper order).

Now, how about the total program? The procedure described in the next-to-last paragraph will make up its main body. Basically, this procedure looks at the contents of memories m_0 through m_7, moving from bottom to top and comparing the numbers two at a time. It moves the smaller number into the higher numbered register and keeps the larger number in the t-register for further comparisons. Go back and read through the procedure to make sure that you understand it. You might want to look at the flowchart in Fig. 2-38, too.

All we have to do to complete the program is to arrange things so that the program runs through the procedure six times and then stops. We can do so by using m_7 (or, it may be more convenient to use ⎡DSZ⎤ on some other register). Start the program with a 6 in m_7. At the end of the main procedure, put in a sequence of steps that successively subtracts 1 from the contents of m_7 and then checks to see if they are down to 0. If they aren't, go back to the top and run the main procedure again. If they are, the procedure has run six times, and therefore you should stop. If you happen to have a TI-57, an extra difficulty must be considered (see Notes). [*Solution was realized on a TI-57 in 37 steps.*]

If you have a calculator with indirect addressing, the above program can be shortened considerably. Notice that in the main procedure you do the same thing over and over (see Fig. 2-38), that is, you keep recalling the contents of a memory and comparing it with the t-register, moving the largest number into the t-register and storing the smaller in a memory. The only things that change are the names of the memories involved. By using indirect addressing (see Sec. 1.9), you can change the names of the memories without having to write down the steps again. You can also shorten the running time by not going all the way to the top on each run through the main procedure. After four runs, for example, the top four numbers are in the proper place; thus, on the fifth run, you need to proceed only as far as m_4 (the fifth memory starting from m_0) and to put the contents of the t-register into it. This method, made possible by indirect addressing, requires the use of a counter that gets incremented (or decremented, depending on how you arrange things) on each pass through the main procedure. What we are getting involved in here is nested loops (see Sec. 1.8). Read the Notes for further remarks on running time.

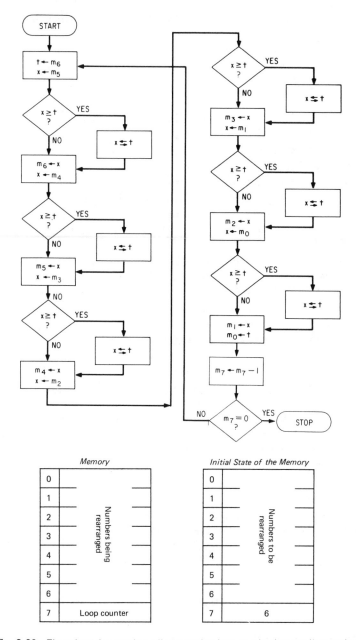

Fig. 2-38 Flowchart for sorting all memories but one in descending order

It should be clear that there is nothing sacred about the number 7 (memories m_0 through m_6) in this problem. The procedures outlined here can be used to sort the contents of *any* collection of memories, limited only by the capacities of your calculator. In particular, if your calculator permits indirect addressing, you should be able to sort the contents of all but a few of its memories (those being reserved for counters and indirect addresses).

Notes: If you happen to have a TI-57, an extra difficulty to consider is this: m_7, the memory you are going to use as a counter, is also the t-register! Consequently, a certain amount of care is called for so that the contents of m_7 do not get lost while tests are being run. The trick is to move the contained number around during the program and keep it out of the way of the rest of the numbers. Let's look at the program to see how this is done. When the program starts, there is a 6 in m_7. The first two steps of the program will be RCL 6 and x $\rightleftharpoons$ t. The next step would ordinarily be RCL 5 to set up the comparison between the contents of m_6 and m_5. However, when x $\rightleftharpoons$ t is executed, the 6 which was in m_7 (alias the t-register) gets pulled into the x-register. If you now execute RCL 5, the 6 will get lost. Therefore, substitute the step EXC 5 instead. This moves the contents of m_5 into the x-register, as you want, but it *also* moves the 6 safely out of the way into m_5. Observe that this maneuver costs you no extra steps. A little later in the program you will want to store a number back in m_5. If you use a simple STO order, the 6 will get lost. Therefore, once again substitute EXC 5. Doing so, of course, pulls the 6 back into the x-register, where it is once again in the way. Therefore, the next step, ordinarily an RCL order, must also be changed to an EXC, and so on. Simply follow the path of the 6 throughout the program, substituting EXCs for STOs and RCLs whenever the 6 threatens to get wiped out. You will find in the end that the contained number can be saved at no cost in program steps!

As we said earlier, a "bubble sort" is not the most efficient algorithm for arranging a list of n numbers in descending order. There is a more efficient method called "sorting by merging." The amount of indirect addressing and loop nesting involved in this method, however, makes it impractical for a calculator.

2.16   The Lagrange Interpolation Formula

Difficulty: 4   Calculator: Variable

Interpolation is a method for approximating the value of a function from other known values. Suppose that you have a table giving the logarithms of the whole numbers 1, 2, 3, $\therefore$., etc., and you want a guess at log(10.82). Our table of logs gives, to four decimals,

$$\log(10) = 1.0000 \qquad \log(11) = 1.0414$$

The question is, can we make a reasonable guess at log(10.82) from this information alone? There is a well-known formula that provides such a guess; it is called the *linear interpolation formula*.

If f is a function and we know the value of f(a) and f(b), where a and b are *different* numbers, this formula says that a reasonable guess for f(c) is

$$(1) \qquad f(c) \cong f(a) \frac{c-b}{a-b} + f(b) \frac{c-a}{b-a}$$

To guess at log(10.82), therefore, we use Eq. (1) with f(x) = log x, a = 10, b = 11, and c = 10.82. Hence,

$$\log(10.82) \cong (1.0000) \frac{10.82 - 11}{10 - 11} + (1.0414) \frac{10.82 - 10}{11 - 10} = 1.0339$$

[Actually, log(10.82) to four decimals is 1.0342.]

Let us see what is behind Eq. (1). Consider the graph of y = f(x) shown in Fig. 2-39. The coordinates of points A and B are known, as is the x-coordinate of point C. We wish to guess at the y-coordinate f(c) of C. To do so, we draw the line ℓ joining A and B and make a vertical line through C, as shown in Fig. 2-40. The point C' where these two lines meet has the same x-coordinate as C, namely c; furthermore, its y-coordinate appears reasonably close to the number f(c). Therefore, we use the y-coordinate of C' as a guess for f(c), and it is this number that Eq. (1) provides.

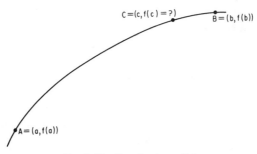

Fig. 2-39 Graph of y = f(x)

To confirm that this reasoning is correct, consider the equation,

$$(2) \qquad y = f(a) \frac{x-b}{a-b} + f(b) \frac{x-a}{b-a}$$

Since it can be put in the form

$$y = mx + t$$

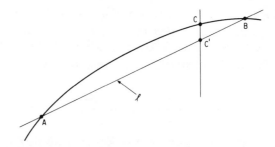

Fig. 2-40   Construction to guess the y coordinate of C

where

$$m = \frac{f(a) - f(b)}{a - b}$$

$$t = \frac{a[f(b)] - b[f(a)]}{a - b}$$

it must be the equation of a straight line in the xy-plane. But note that if $x = a$ in Eq. (2), then $y = f(a)$, and if $x = b$, then $y = f(b)$. Consequently, this straight line must pass through the points [a, f(a)], or A, and [b, f(b)], or B, and hence must be the same as line ℓ.

Since point C' is on line ℓ, its y-coordinate can be obtained by setting x equal to c in Eq. (2), thus yielding Eq. (1). We will give no solution for the following problem; you are strictly on your own.

Problem 1:   Write a program that accepts as inputs the numbers a, f(a), b f(b), and c, and outputs the guess f(c) obtained by Eq. (1).

Our table of logs also reveals that $\log(12) = 1.0792$. Can this additional information help us get a better guess for log (10.82)? Yes, it can. There happens to be another interpolation formula that can be used to guess the value of f(c) if we know *three* different values, $f(x_1)$, $f(x_2)$, and $f(x_3)$. A reasonable guess for f(c) is then

$$(3) \quad f(c) = f(x_1)\frac{(c - x_2)(c - x_3)}{(x_1 - x_2)(x_1 - x_3)} + f(x_2)\frac{(c - x_1)(c - x_3)}{(x_2 - x_1)(x_2 - x_3)}$$

$$+ f(x_3)\frac{(c - x_1)(c - x_2)}{(x_3 - x_1)(x_3 - x_2)}$$

Using this formula with $x_1 = 10$, $f(x_1) = 1$, $x_2 = 11$, $f(x_2) = 1.0414$, $x_3 = 12$, $f(x_3) = 1.0792$, and $c = 10.82$ gives

$$\log(10.82) \cong 1.0342$$

a figure that is accurate to four decimals.

Let us examine the basis for this guess. The equation

$$y = f(x_1) \frac{(x - x_2)(x - x_3)}{(x_1 - x_2)(x_1 - x_3)} + f(x_2) \frac{(x - x_1)(x - x_3)}{(x_2 - x_1)(x_2 - x_3)}$$
$$+ f(x_3) \frac{(x - x_1)(x - x_2)}{(x_3 - x_1)(x_3 - x_2)}$$

can be in the form: $y = mx^2 + nx + t$. Its graph, therefore, is a parabola. Note also that when x is x_1, x_2, or x_3, then y is $f(x_1)$, $f(x_2)$, or $f(x_3)$, respectively. Consequently, just as Eq. (1) makes a guess by passing a straight line through two given points, Eq. (3) makes a guess by passing a parabola through three given points. And, by the way, just as any two points determine one and only one line, any three points (in this case) determine one and only one parabola (or only one line, if m = 0).

The general formula, called *LaGrange's Interpolation Formula*, is stated as follows: If x_1, x_2, . . ., x_n are n different numbers and if $f(x_1) = y_1$, $f(x_2) = y_2$, . . ., $f(x_n) = y_n$, then a reasonable guess at $f(x)$ is given by

$$y = y_1 \left(\frac{x - x_2}{x_1 - x_2} \right) \left(\frac{x - x_3}{x_1 - x_3} \right) (\cdot \cdot \cdot) \left(\frac{x - x_n}{x_1 - x_n} \right)$$

(4)
$$+ y_2 \left(\frac{x - x_1}{x_2 - x_1} \right) \left(\frac{x - x_3}{x_2 - x_3} \right) (\cdot \cdot \cdot) \left(\frac{x - x_n}{x_2 - x_n} \right) + \cdot \cdot \cdot$$

$$+ y_n \left(\frac{x - x_1}{x_n - x_1} \right) \left(\frac{x - x_2}{x_n - x_2} \right) (\cdot \cdot \cdot) \left(\frac{x - x_{n-1}}{x_n - x_{n-1}} \right)$$

$$= \sum_{i=1}^{n} y_i \prod_{\substack{j=1 \\ j \neq i}}^{n} \frac{x - x_i}{x_i - x_j}$$

(See Sec. 1.8 for an explanation of the symbols Σ and Π.)

Equation (4) is of the form

(5) $$y = a_n x^n + a_{n-1} x^{n-1} + \ldots + a_1 x + a_0$$

where a_n, a_{n-1}, . . ., a_1, a_0 are complicated expressions involving the known numbers x_1, x_2, . . ., x_n, y_1, y_2, . . ., y_n. Equation (5) is called a *polynomial* of degree n (or less, if $a_n = 0$). Note also that setting x = x_i in Eq. (4) gives $y = y_i$ for each i = 1, 2, . . ., n; Eq. (4), therefore, gives the equation of the (one and only) polynomial of degree n or less whose graph passes through all of the points (x_i, y_i), where i = 1, 2, . . ., n.

Problem 2: Write a program that will accept as inputs the numbers x_1, x_2, . . ., x_n, y_1, y_2, . . ., y_n, and x and that will output the number y given by Eq. (4).

Naturally, since the number of storage registers your calculator has will limit the number n, you must choose the largest n available. [Our HP 67/97 program will handle an n as large as 9; a TI 58 or 59 program should be able to handle one much larger. You should be able to handle an n of 11 or so on on an HP 19/29c, and perhaps an n of 3 on an HP25(c)]. For any n much larger than 4, you will probably need indirect addressing (see Sec. 1.9).

Solution: The program requires many registers and indirect addressing (see Sec. 1.9). Let us store the numbers x_1, x_2, . . ., x_n, y_1, y_2, . . ., y_n in registers r_1, r_2, . . ., r_n, s_1, s_2, . . ., s_n, respectively. We will also need eight other registers, as follows:

m_0	Register to accumulate Σ	m_4	For storing x_i temporarily
m_1	Register to accumulate Π	m_5	For storing x
m_2	Counter for i	m_6	For storing n
m_3	Counter for j	I	Indirect register

Thus, a total of $(2n + 8)$ registers are needed. (If you are desperate for storage registers, you can store m_2, m_3, and m_6 in one register, but this will increase the number of program steps—see Sec. 1.14.)

The program (or the user) should begin by storing x_1, x_2, . . ., x_n, y_1, y_2, . . ., y_n, x, and n in registers r_1, r_2, . . ., r_n, s_1, s_2, . . ., s_n, m_5, and m_6, respectively (see Sec. 1.5). The following mess must now be evaluated:

$$y = \sum_{i=1}^{n} y_i \prod_{\substack{j=1 \\ j \neq i}}^{n} \frac{x - x_j}{x_i - x_j}$$

This is, of course, the sum of the n expressions

$$t_1 = y_1 \prod_{j=2}^{n} \left(\frac{x - x_j}{x_1 - x_j} \right), t_2 = y_2 \prod_{\substack{j=1 \\ j \neq 2}}^{n} \left(\frac{x - x_j}{x_2 - x_j} \right), \ldots,$$

$$t_i = y_i \prod_{\substack{j=1 \\ j \neq i}}^{n} \left(\frac{x - x_j}{x_i - x_j} \right), \ldots, t_n = y_n \prod_{j=1}^{n-1} \left(\frac{x - x_j}{x_n - x_j} \right)$$

Let us look at one of these terms, say

$$t_i = y_i \prod_{\substack{j=1 \\ j \neq i}}^{n} \left(\frac{x - x_j}{x_i - x_j} \right)$$

and see how to evaluate it.

Consider the flowchart in Fig. 2-41, which begins with i in register m_2. This flowchart calculates t_i, using the techniques described on pp. 66-68. Register m_3 stores j, beginning at the value, j = n, and decrementing by 1 each step through the loop. Register m_1 accumulates the product t_i. It begins at $m_3 = y_i$ and successively accumulates the values

$$m_3 = y_i \left(\frac{x - x_n}{x_i - x_1} \right), \; m_3 = y_i \left(\frac{x - x_n}{x_i - x_n} \right) \left(\frac{x - x_{n-1}}{x_i - x_{n-1}} \right),$$

$$m_3 = y_i \left(\frac{x - x_n}{x_i - x_n} \right) \left(\frac{x - x_{n-1}}{x_i - x_{n-1}} \right) \left(\frac{x - x_{n-2}}{x_i - x_{n-2}} \right) \; \cdots$$

except that the factor

$$\frac{x - x_i}{x_i - x_i} = \frac{x - x_i}{0}$$

is to be skipped.

The first box in the flowchart initializes registers m_3 and m_1, as described above, and also stores x_i in m_4. The test, $m_4 = r_{m_3}$?, will be answered "yes" only when $r_{m_3} = x_j = x_i$ or when i = j since the numbers $x_1, x_2, \ldots, x_n$ are all different. Thus this test will omit the undefined factor

$$\frac{x - x_i}{x_i - x_i}$$

If this test is answered "no," then i $\neq$ j, and the next box updates m_1.

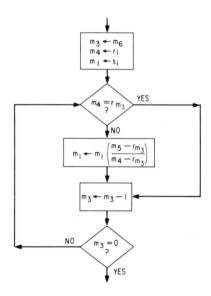

Fig. 2-41 Flowchart for calculating t_i

The next box decrements the register, $m_3 = j$, for the next run through the loop, and the last test, $m_3 = 0$?, stops the program after n runs through the loop. On only one of these n runs (when $i = j$) is the m_1 register not updated, and that is at the factor to be skipped. Hence the routine terminates with

$$m_1 = y_i \prod_{\substack{j=1 \\ j \neq i}}^{n} \frac{x - x_j}{x_i - x_j} = t_i$$

We now proceed to evaluate the Σ. To do so, consider the flow-chart in Fig. 2-42. This flowchart calculates the required number y, again using the techniques described on page 000. Register m_2 stores i, beginning at $i = n$ and decrementing by 1 on each run through the loop. Register m_0 accumulates the sum, $y = t_1 + t_2 + \ldots + t_n$, beginning at $m_0 = 0$ and successively accumulating the numbers, $m_0 = t_n$, $m_0 = t_n + t_{n-1}$, $m_0 = t_n + t_{n-1} + t_{n-2}$, etc.

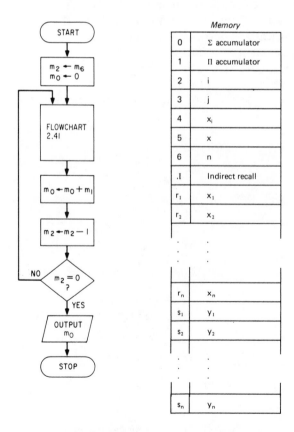

Fig. 2-42 Flowchart for solution to Problem Z

The first box initializes registers m_2 and m_0. On each run through the loop, the flowchart in Fig. 2-40 calculates the corresponding t_i. The next box, $m_0 \leftarrow m_0 + m_1$, updates m_0. The next box decrements m_2 for the next run through the loop, and the test, $m_2 = 0?$, derails the loop after n runs through the loop. Since the answer we want, y, is now in m_0 ($m_0 = t_1 + t_2 + \ldots + t_n = y$) we output m_0 and stop. [*Solution was realized on an HP-67 in 37 steps for $n \leqslant 9$.*]

2.17   Random Walks

Difficulty: 2

As an example of the *random walk*, consider that Henry and Ted flip coins for cigarettes. Every time the coin lands heads up, Henry wins a cigarette from Ted; every time tails turns up, Ted wins a cigarette from Henry. Each begins the game with a full pack of twenty cigarettes, and the game continues until one player wins all forty. Other names for the random walk range from the *Markov Process* to *drunkard's walk*. Loosely, a Markov Process can be thought of as a kind of machine that can be in one of several different states. It changes from state to state according to certain probabilities, and sometimes it just stops dead.

Problem 1:   Write a program to simulate the above game. It should keep track of how many cigarettes Henry and Ted have at each step. It should flip a fair coin (see Sec. 3.2), display the outcome, make the appropriate transfer of one cigarette, and display the number of cigarettes each player has at that point. It should then test to see whether either player has been busted, and if not, it should go back and flip the coin again.

In this game, for which no solution will be given, the number of cigarettes in Henry's possession at any time, h, completely determines the state of the game, for Ted must then have $(40 - h)$ cigarettes. Thus, there are exactly 41 possible states, corresponding to h = 0, 1, 2, ..., 39, 40. At states in which h = 0 or h = 40, the game ends, since, if h = 0, Henry's busted, and if h = 40, Ted is. In the other 39 states (h = 1, 2, ..., 39), the game changes either to state $(h + 1)$—Henry wins a cigarette—or state $(h - 1)$—Ted wins a cigarette. Since presumably a fair coin is being tossed, there's a 50-50 chance for either alternative.

Problem 2:   A drunkard is staggering up and down the length of a 10-foot plank floating in the middle of a swimming pool. Every 5 seconds, he does one of three things:

1. Staggers 1 foot to the right
2. Staggers 1 foot to the left
3. Stands perfectly still, in a stupor

He chooses from these alternatives with equal probability, indulging in each one-third of the time, until finally he staggers off one end of the plank or the other and into the water.

To simulate the drunkard's walk on your calculator, let the 10-foot plank be represented by ten 1's in your display, as follows:

$$1111111111$$

and let the drunkard be represented by the digit 8. Thus, the state in which the drunkard is 3 feet from the left end of the plank would be represented by the number 1181111111. The program should start the drunkard in one of the two center positions 1111181111 or 1111811111 (or, if desired, at a position chosen by the user).

The whole program will constitute a loop. On each pass through this loop, one of the three alternatives should be chosen, the position of the 8 in the plank of 1's adjusted accordingly, a test made to see whether or not the drunkard has fallen off the end, and if not, the new state should be displayed and the run repeated.

Solution: Since each stage involves a random choice among three alternatives, we must use a randomizer. It is convenient to have a random number generator that outputs a random integer c between -1 and 1, inclusive (see Problem 4 of Sec. 1.11 for a way in which to execute this). If the randomizer outputs a c of -1, we move the drunkard one step to the right; if $c = 0$, he stands still; and if $c = 1$, he moves one step to the left. We will use register m_0 to store a random seed and the successive random numbers in the range, $0 \leq m_0 \leq 1$, as described in Sec. 1.11.

Our program stores the empty plank, 1111111111, in register m_1. [This number can be easily calculated by noting that 1111111111 $= 1/9(10^{10} - 1)$.] To put the drunkard (the digit 8) in the appropriate position on the plank, we must add the digit 7 to one of the 1's on the plank. For example, to place the drunkard in the third place from the right, we have to add 700 to the plank since 1111111111 $+$ 700 $=$ 1111111811. Thus, we will store the current position of the drunkard in m_2. This register will contain one of the ten numbers, 7000000000, 700000000, . . ., 700, 70, 7, depending on where the drunkard is located on the plank. Thus, if the drunkard is at the extreme left of the plank, $m_2 = 7000000000$, if he is on the third place from the right, $m_2 = 700$. Note that regular m_2 serves a dual purpose. It keeps track of the drun-

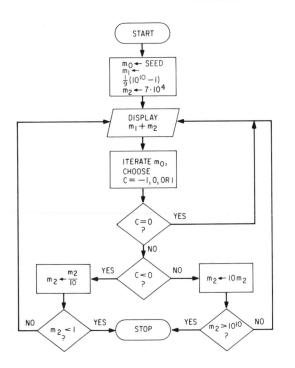

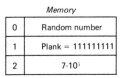

Fig. 2-43 Flowchart for the drunkard's random walk

kard's position and also serves as the number to be added to the plank, $m_1 = 1111111111$, to produce the desired output.

To move the drunkard, we first test to see if $c = 0$. If it is, we proceed directly to output, as the poor guy is standing stock still. If $c \neq 0$, we next test to see if $c < 0$. If so, $c = -1$, and we must move the drunkard one step to the right (that is, we divide m_2 by 10). If not, $c = 1$, and we must move the drunkard one step to the left (by multiplying m_2 by 10). In each of the last two cases, we must make sure that the drunkard has not fallen off the plank before we proceed. In the case, $c = -1$, the drunkard will have just fallen off the right end if $m_2 = 0.7$ (we can check this contingency by using the test, $m_2 < 1$). If $c = 1$, the drunkard will have fallen off the left end if $m_2 = 7 \times 10^{10}$ (we can check this contingency by using the test, $m_2 > 10^{10}$). The program stops if he has fallen off either end. Otherwise, it pauses to display the output, $m_1 + m_2$, as discussed above, and starts all over again. The flowchart is shown in Fig. 2-43. [*Solution was realized on an HP-25 in 49 steps.*]

Reliably, at any nighttime moment (that is, nonbusiness hours) in North America hundreds of computer technicians are effectively out of their bodies, computer-projected onto cathode ray tube display screens, locked in life-or-death space combat for hours at a time, ruining their eyes, numbing their fingers in frenzied mashing of control buttons, joyously slaying their friends and wasting their employers' valuable computer time. Something basic is going on.

STEWART BRAND
Fanatic Life and Symbolic Death Among the Computer Bums

CHAPTER THREE

Games

3.1 Introduction

Although computers (and programmable calculators) were not invented for the purpose of playing games, it didn't take programmers long to figure out that the two are admirably suited to one another. A visit to any college computing center will bear out this observation. In addition, many of the rapidly multiplying electronic games on the market today have at their heart a microprocessor especially designed for their use alone. The games in bars and game arcades as well as those played at home on a television set fall into this category.

In order for a game (or anything else for that matter) to be put onto a programmable calculator, it must admit a description in terms of numbers and the workings of the game must be capable of being "modeled" in terms of the operations available on the calculator. Many games are open to this kind of treatment in whole or in part. There are, in fact, several specific aspects of games that can be "calculatorized."

One such, featured prominently in this chapter, is game equipment: the boards, cards, dice, spinners, and other paraphernalia with

197

which games are played. It is easy, for example, to see that the "output" of a pair of dice or a roulette wheel is representable by numbers, for this output is already in numerical form in its "natural" state. The only difficulty is how to randomize the numbers and keep them in the proper range.

Most of the problems in this chapter involve some modeling of game equipment. They range in difficulty from No. 3.2, which asks you to simulate the tossing of a coin, to No. 3.17, in which the calculator takes over a function performed by the player in traditional forms of the game. In all these problems, the game is translated from the real world onto the calculator without undergoing any essential change. The equipment with which the game is played, however, is fairly radically altered, for in each case it is replaced by the calculator. Technically, the original equipment is *simulated* by the calculator.

Now, it is also possible to simulate entire games on the calculator. In such cases, the game as a whole is radically altered by being translated out of the real world into the world of numbers on the calculator. We provide two examples of this type of game, Problems 3.12 and 3.13 (hockey and basketball). In these, the physical setting of the game is translated onto the calculator by means of a set of coordinates. In hockey, for example, the playing surface is replaced by an x, y coordinate plane; the puck becomes a point in that plane; and the goal is an interval on the x-axis. The skills required to play the game also change, in this case from agility, stamina, and reflexes to the ability to judge angles (numerically, of course). Such games are related to the many electronic games in which a sport such as tennis or skeet shooting is simulated on a cathode-ray tube.

Besides fulfilling the passive role of equipment, the calculator can also take part as a player. Undoubtedly, the most impressive computer programs of this type are those for playing chess since computers can now play it at the Master level! Of course, one couldn't hope to get a strategy for such a complicated game onto a calculator, but several fairly simple games for which a winning strategy exists can be taught to the calculator. We provide four of them in this chapter, Problems 3.5, 3.14, 3.15, and 3.16. Problem 3.5 is coupled with Problem 3.4 in an interesting way; they represent two sides of a guessing game. In Problem 3.4, the calculator acts as the game equipment, returning clues as the user makes guesses. In Problem 3.5, the calculator makes the guesses and the user returns the clues. Problems 3.14, 3.15, and 3.16 are games in which the rules for a winning strategy are to be programmed onto the calculator. Although all of the strategies are simple enough to be put onto a small calculator (with squeezing in some cases), none is obvious. If the player makes a single mistake, the calculator wins.

Finally, we must say a word about random numbers, which play a part in almost half the games in this chapter. One of the major divisions in the science of game theory is between games that involve random events and those that do not. All "games of chance" fall into the first category because they rely on such random events as rolling a pair of dice, drawing from a deck of shuffled cards, or spinning a wheel. Problems 3.2, 3.3, 3.4, 3.6, 3.7, and 3.8 all involve the element of randomness.

In order to simulate random events on your calculator, you will need to learn how to use a random number generator [we should really say *pseudo*-random number generator, since no process that uses a well-defined rule for producing a string of numbers can be truly random (see Problem 3.9)]. Consult your owner's manual or refer to Sec. 1.11 for techniques. We will use the notation, $x \leftarrow \text{Rand}(m_0)$, to mean "generate a random number between 0 and 1 from the random number in m_0." This step will be the heart of each of the random-number-generating routines required here. Your calculator may have its own random number generator built in. If so, you can read the notation, $x \leftarrow \text{Rand}(m_0)$, to mean "generate a random number between 0 and 1 using your random number generator (and place it in the display)." The problems have been arranged in ascending order of difficulty in the use of such a generator so that if you've never worked with one before, you can gradually acquire the needed skills.

Randomness can also be part of the *strategy* for playing a game; for instance, the strategy of bluffing in poker must be somewhat random if it is to be effective. We have not included an example of a strategy of this kind in this collection of problems, but perhaps some clever reader will be able to fill the gap.

3.2   Calculator Coin Toss

Difficulty: 1

In order to model games involving random events, you will need to be able to generate all kinds of random numbers. Your calculator's applications manual will provide a program for generating a sequence of evenly distributed random numbers between 0 and 1.[1] This will serve as your basic tool for generating other sequences of random numbers.

Problem: Write a program that simulates tossing a coin, letting 0 stand for tails and 1 for heads. Each run of the program should return randomly a 0 or a 1.

---

[1] If not, see Sec. 1.11.

Solution: The idea is to separate the random numbers into two classes, assigning a 0 to one class and a 1 to the other. You can do so in many ways, but perhaps the easiest is to multiply the random number between 0 and 1 by 2. The result will be a random number between 0 and 2. If the number is less than 1, return a 0. If it is greater than or equal to 1, return a 1. If your calculator has an "integer part" key, this operation can be accomplished neatly by taking just the integer part of the number between 0 and 2. The flowchart is shown in Fig. 3-1. [*Solution was realized on an HP-25 in 11 steps.*]

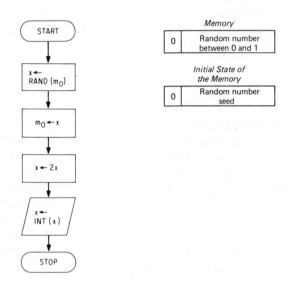

Fig. 3-1 Flowchart for tossing a coin

Note: It might seem that the preceding program is slightly weighted in favor of heads, since the "dividing line," 1, is included in the heads pile. The difference this makes is miniscule, of course, and can be eliminated by not returning anything and going back for another number when a 1 turns up as the random number between 0 and 2. However, many random number generators produce a random number between 0 and 1 by taking the fractional part of some number. (This is true of the random number generators described in Sec. 1.11, for example). Since the number so generated *can* be 0 but *cannot* be 1, it is slightly weighted toward 0. This effect is exactly balanced by throwing the "dividing line" into the heads pile. You might analyze the random number generator for your calculator from this point of view.

3.3   Dice Roll

Difficulty: 1

Problem: Write a program to simulate a pair of dice, in which each run produces a *pair* of random whole numbers between 1 and 6 (inclusive). Arrange things so that the output will be a two-digit number, with one digit preceding and the other following the decimal point, as follows: a.b ("a" represents the number on the first die and "b" the number on the second die).

Solution: The basic routine for solving this problem is to (1) generate a random number between 1 and 6 and store it somewhere; (2) generate a second random number between 1 and 6, multiply it by .1, and add it to the first number; and (3) return with the result. Since you are going to generate a random number between 1 and 6 twice, it would be inelegant not to use the same set of program steps both times. This means, however, that a way of knowing which number has just been generated is required so that you will know whether to multiply it by .1 or not. To this end, start with 0 in the memory that is being used to store the first number. Check the contents of the memory when you get to the end of the random-number generating routine. If it is 0, store the number you have just generated. If it is not, then the number you have generated is the second and must be operated on in the manner specified. The program will need two memories:

m_0:   Random number seed

m_1:   Temporary storage for the number on the first die (and perhaps some other data; see discussion below)

The program then proceeds as follows: Recall the random number seed and use it to generate a new random number (and new seed) between 0 and 1. Multiply the number by 6 and add 1, thereby yielding a random number between 1 and 7. Now "round down" by taking the integer part, the result being a random whole number between 1 and 6. At this point, since the calculator doesn't know whether it is on its first or second run through the random number generating routine, recall the contents of m_1. If it is zero, store your random number—which you now know is the number on the first die—in m_1 (be careful not to lose it in the testing procedure!); then go back to the beginning of the program for another random number. If m_1 is not zero, the first die is already in m_1. Consequently, multiply the present random whole number by .1, add it to the contents of m_1, and return with the result in the display. Before quitting, erase memory m_1 to be ready for the next run. The flowchart is shown in Fig. 3-2. [*Solution was realized on an HP-25 in 26 steps.*]

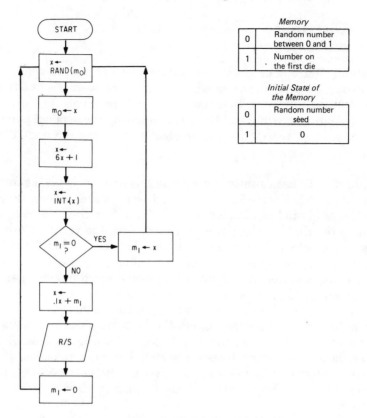

Memory	
0	Random number between 0 and 1
1	Number on the first die

Initial State of the Memory	
0	Random number seed
1	0

Fig. 3-2 Flowchart for dice roll

Note: This last sequence of events can be accomplished neatly with register arithmetic. If you have just multiplied the number on your second die by .1 and are ready to add it to the contents of m_1, the rest of the program can proceed as follows:

$$\begin{array}{l}
\text{TI} \\
\text{SUM} \\
01 \\
\text{RCL} \\
01 \\
\text{INV} \\
\text{SUM} \quad \text{(erases } m_1\text{)} \\
01 \\
\text{HP} \\
\text{STO} + 1 \\
\text{RCL 1} \\
\text{STO} - 1 \quad \text{(erases } m_1\text{)}
\end{array}$$

Note that the previous contents of m_1 remain in the display.

If your calculator has subroutines, the part of the program that generates a random whole number between 1 and 6 should be a subroutine. This will obviate the need for the $m_1 = 0$ test.

3.4   HI-LO I

Difficulty: 2

Here is a game sometimes played on the radio, where it is called *high-low*. The player is supposed to guess a (whole) number between some given bounds. After each guess, the player is told whether his guess was too high or too low (or correct). On the radio you get only one guess, but here on your own calculator, you can program the game any way you want. Let's not make it as hard as it is on the airwaves.

Problem: Write a program for the calculator to play the following game: At a given signal, the calculator is to generate a random whole number between 1 and n (n is a variable that will be stored in some memory location; you can set it to any value, depending on how hard you want to make the game). Once the random number is hidden in a memory, the game begins. The player enters a series of guesses. After each guess, the calculator returns a 1 if it was too large, a -1 if too small, and a 0 if correct. The calculator should also keep score so that once the correct guess is entered, the calculator returns (after the 0) the number of guesses it took the player to get the right answer.

Solution: The program has to do two separate things: (1) generate and store the random number between 1 and n, and (2) evaluate guesses at the number by the player and keep score. If your calculator has user-defined keys, you can use different keys to handle these separate functions. If not, you will have to use some kind of signal to enable the calculator to execute the part of the program you want. For example, you might use 0 as a signal that you want the calculator to execute Part (1) of the program. Any other number would then be interpreted as a guess at the number already stored in m_1. Thus, the first step in the program would be to look at the number in the display to see if it is a 0 or not. If it is, branch to Part (1). If not, treat the number as a guess (don't lose it in the testing process!) and branch to Part (2).

Here are the numbers that will be needed in memory:

m_0 Random number seed
m_1 Number generated in Part (1) that the player is trying to find

m_2 Scorecard that counts the number of guesses taken so far by the player

m_3 n, the "upper bound" on the number in m_1

If you have done the previous two problems, you should have no trouble with Part (1) of this program. The only wrinkle here is that the range of random numbers to be generated is supposed to be controllable by the player. Having stored the upper bound on the range of numbers to be produced in m_3, the player can select whatever range he wants by storing the appropriate number in m_3 before the program is run. The number in m_3 then plays the role of the multiplier, 6, in the solution to the previous problem. The only other thing required in this part of the program, after generating the number in m_1, is to erase the scorecard (that is, store a 0 in m_2).

Part (2) begins when 1 is added to m_2 (don't lose the guess). Next recall m_1 and start comparing it with the guess to see if the latter was too large, too small, or correct. There are two ways that you can make the comparisons: by comparing the guess directly with m_1 or by subtracting m_1 from the guess and checking to see if the result is greater than, equal to, or less than 0. Which of these is more efficient will depend on what type of logic your calculator has and what its testing capabilities are. In either case, you will need two successive tests to sort out whether the guess was too large, too small, or correct. For example, you might first test whether m_1 is greater than or equal to the guess. If it isn't, the guess is definitely too large. If it is, you will then have to separate "greater than" from "equal to." If the guess was too large or too small, return a plus or minus 1 accordingly and "reset" the calculator for the next guess. If the guess was correct, return a 0 for two beats, recall the number in m_2, and reset (resetting will not be necessary if you are using labels). The flowchart is shown in Fig. 3-3. [*Solution was realized on an HP-25 in 38 steps.*]

3.5   HI-LO II
Difficulty: 2

In this game, the problem is the same as for the previous one except that the roles of the calculator and the player are reversed. This time after the number n (see previous problem) has been stored in a memory, the player will generate and store a number between 1 and n in his own memory. It will then be the calculator's job to find it. Each time the calculator makes a guess, the player will enter a 1, -1, or 0 indicating that the guess was too large, too small, or correct. With this number as "feedback," the calculator makes its next guess, or, if its last guess was correct, returns the number of guesses it took to get the right answer.

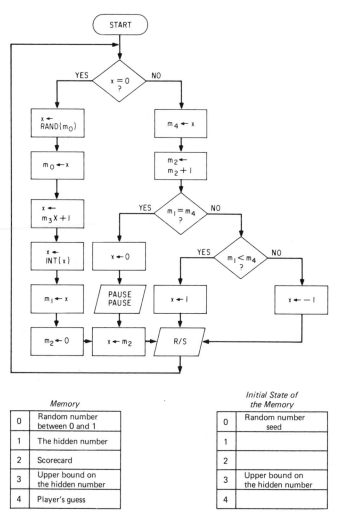

Fig. 3-3 Flowchart for HI-LO I

There are various ways in which the calculator could make its search for the correct number. One would be for it to start at 1 and keep increasing its guess by 1 until the correct number is reached. Obviously, this is not the best possible strategy. Notice that it makes no use of the feedback generated by previous guesses. If you played the previous game with your calculator, you undoubtedly used feedback from previous guesses to "zero in" on the correct answer. The calculator can be programmed to do the same thing. If you or the calculator are using the optimal strategy, it should never take more than log(n)/log(2) guesses, rounded up to the next larger whole number, to get the right answer.

Solution: We will start our explanation of the solution in the middle and work toward both ends. How does the calculator make its guesses? It remembers its last too small guess and its last too large guess and makes its next guess exactly halfway between them (that is, the average). A small problem with this strategy is that it doesn't always output a whole number, but this difficulty can be taken care of by setting the display to zero decimal places. Then, although the calculator's guesses will cease to be whole numbers *internally*, the output in the display will always be in the correct form. The memories for the program will contain the following:

m_1 Last guess that was too small
m_2 Last guess that was too large
m_3 Present guess (average of m_1 and m_2)
m_4 Scorecard (records how many guesses have been made so far)
m_5 n, the upper bound on all guesses

Suppose that the calculator has made a guess and stored it in m_3. How will it make its next guess? The player will have entered a 1 or -1 (we'll worry about 0 later), indicating that the guess in m_3 is too large or too small. If too large, it then becomes the last guess that was too large, and its contents are moved into m_2. The new m_1 and m_2 are then used to make the next guess, which once again is stored in m_3. Thus, the middle part of the program goes like this: Assuming that a 1 or -1 has been entered, the contents of m_3 are recalled and stored in m_2 or m_1, accordingly. Then the average of m_1 and m_2 is computed and stored in m_3. The program returns with this value in the display. Somewhere in this part of the program, moreover, a 1 should be added to the scorecard.

Now how does the calculator make its *first* guess? Since the logical first guess is n/2, we should have the memories set so that the calculator will make this guess at the beginning. We will also want to get a 0 into m_1 and an n into m_2. These are the initial "bounds on the solution." A little reflection will convince you that the following initial configuration of the memories will get the calculator to make the right first guess if a 1 is entered in the display:

m_1 0
m_2 0
m_3 n
m_4 0
m_5 n

Thus, the calculator should be started in this configuration. After the player, by entering a 1, gets the calculator to make its first guess, the memories will look like this:

m_1	0
m_2	n
m_3	n/2
m_4	1
m_5	n

Finally, the entry of a 0 indicates that the calculator's present guess is correct. In this case, the correct guess should be recalled and displayed for three beats. Next, the memories should be reset into the "initial configuration" already discussed, to be ready for the next round. Last, the contents of m_4 should be returned and shown in the display (don't erase m_4 before retrieving this number!). Entering a 0 can also be used to initialize the memories for the first round of the game, once n has been stored in m_5. The flowchart is shown in Fig. 3-4. [*Solution was realized on an HP-25 in 30 steps.*]

Note: The foregoing program is an example of a *binary search,* in which the number being sought is trapped between a pair of numbers that keep getting closer to one another as the program proceeds and at each stage the search area is cut in half (hence the name for the

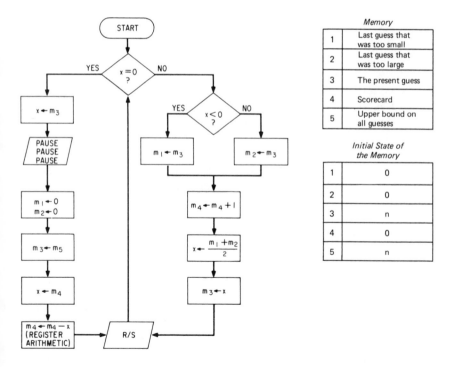

Fig. 3-4 Flowchart for HI-LO II

technique). This is an old and venerable mathematical trick still in widespread use today, especially in connection with computers. The earliest known example is the Babylonian algorithm for computing square roots, an algorithm dating back to at least 1800 B.C.

It works as follows: Let x be the number whose square root is to be found. To get the algorithm going, one must first guess an answer. The guess does not have to be a good one, just something for the algorithm to work on. Let the first guess be labelled x_0. Suppose, for the purpose of simplifying the explanation, that x_0 is too large (although this is not necessary, as the algorithm works for any case). For example, if $x > 1$, you might let $x_0 = x$. Now if you let $y_0 = x/x_0$, then $x_0y_0 = x$. Moreover, since $x_0 > \sqrt{x}$, we must have $y_0 < \sqrt{x}$ (otherwise $x_0y_0 > x$). Thus $\sqrt{x}$ is trapped between y_0 and x_0, as follows: $y_0 < \sqrt{x} < x_0$.

The algorithm is now ready to start generating its own guesses. At each stage, the search area is "cut in half" by making a guess exactly halfway between the previous two "bounds on the solution" (at this stage, x_0 and y_0). Hence, the second guess (the *calculator's* first) will be the following: $x_1 = (x_0 + y_0)/2$. As it turns out, this guess is always too large since, in every case, $\sqrt{x} < x_1$.[2] Let $y_1 = x/x_1$. Then, as above, $x_1y_1 = x$ and $y_1 < \sqrt{x} < x_1$. Moreover, $y_0 < y_1$ (since $x_0y_0 = x_1y_1$ and $x_1 < x_0$). Thus, we have $y_0 < y_1 < \sqrt{x} < x_1 < x_0$. At this point, $\sqrt{x}$ is trapped between two numbers whose distance from one another is less than half the distance between x_0 and y_0. The algorithm continues to make guesses using the same scheme: $x_3 = (x_1 + y_1)/2$; $y_3 = x/x_3$; $x_4 = (x_3 + y_3)/2$; etc. At each stage, $\sqrt{x}$ will continue to get trapped between a pair of numbers whose distance from one another is less than half the distance between the previous two.

Thus, the numbers x_1, x_2, x_3, . . . (also y_1, y_2, y_3) converge very rapidly toward $\sqrt{x}$. Of course, your calculator has its own square root function, but you might enjoy programming this one on your calculator to see how efficient it is. See Sec. 1.15 for a further discussion of binary searches.

3.6   Roulette

Difficulty: 2
Calculator size: Small or medium

Another game that can be simulated with a random number generator is roulette. An American roulette wheel has 38 markers on it: the numbers 1 through 36, 0 and 00. The wheel and a little ball are

---

[2] Another venerable fact is that $\sqrt{x} = \sqrt{x_0y_0} < (x_0 + y_0)/2$. This inequality, whose technical statement is that the geometric mean of two numbers is smaller than the arithmetic mean, has been attributed to Pythagoras.

set in motion in opposite directions, the ball eventually coming to rest on one of the markers. Various bets can be placed on where the ball will land. The bet with the highest return (and the lowest chance of paying off) is one made on the exact number of this landing spot. A correct guess returns 37 times the original bet (including that bet). The bet with the lowest return (and the highest chance of paying off) is one made on the ball's landing on an odd or even number. As these numbers are colored red and black, respectively,[3] this bet is referred to as a bet "on red" or "on black" (since 0 and 00 are colored green, a red-black bet will be lost if either of these numbers comes up). A winning red-black bet pays back twice the original bet.

Although other bets are possible, let's forget about them for the moment and consider a problem that can be programmed on a small-capacity calculator (such as an HP-33E).

Problem: Write a program that takes bets on a particular number and on red or black. The user is to put (1) the amount bet on a number, (2) the number bet on, (3) the amount bet on a color, and (4) the colors red and black (let 1 stand for red and 0 for black) into four different memories. When the program is run, it first "spins the wheel" and then returns (1) the number on the wheel, (2) the payoff on the number bet, and (3) the payoff on the red-black bet.

If your calculator has user-defined keys, you can produce a more automated program by using some of them to take bets. For example, you might use key A to record a bet on red or black and key A' to record the amount bet. Since a calculator with user-defined keys can probably handle a larger program as well, you might consider adding some other bets. Two other common ones are as follows:

1. *High or low,* in which the bet is on the ball's landing between 19 and 36 (high) or between 1 and 18 (low). There is no payoff for 0 and 00, but for a winning bet the payoff is 2 for 1.
2. *Dozens,* in which the bet is on the ball's landing in the first (1–12), second (13–24), or third (25–36) dozen numbers (once again, 0 and 00 do not pay off). The payoff here on a winning bet is 3 for 1.

Solution: First let's label the memories that will record the bets as follows:

m_1 Number bet on
m_2 Amount bet on the number

---

[3] Not being roulette players, the authors did not duplicate the roulette wheel exactly. In actuality, both odd and even numbers can be black or red.

m_3 Color red or black (1 or 0, respectively)

m_4 Amount bet on red or black

In addition, four other memories will be used, labeled as follows:

m_5 Random number seed (a number between 0 and 1)

m_6 36

m_7 38

m_8 number that comes up on the wheel

The numbers in m_6 and m_7 are constants used by the program.

The program can be broken into three parts: (1) spinning the wheel, (2) checking the bets and paying off the winners, and (3) displaying the results. Since 00 and 0 have to be distinguished from one another, let -1 stand for 00. Then spinning the wheel is just a matter of generating a random whole number between -1 and 36 (inclusive). If you have done some of the previous problems, you should have no trouble doing this: Just use the contents of m_5 to generate a random number between 0 and 1 (don't forget to store the new number in m_5), multiply the result by 38 (the contents of m_7), take the integer part of the result and subtract 1, and then store the result in m_8. This completes the first part of the solution.

Now for the second part, the payoff on the winning bets. Recall the contents of m_1 to see if they are equal to the contents of m_8. If they are, the number bet wins, and the contents of m_2 should be multiplied by 37 (leaving the result in m_2). If not, the bet is collected by setting the contents of m_2 equal to 0. Next, check to see if the contents of m_8 are ≤ 0 (that is, if m_8 is either 00 or 0). If so, the red-black bet loses. Set the contents of m_4 equal to 0 (collect the bet), and skip directly to the third part of the solution. If $m_8 > 0$, you must determine whether it is odd or even (red or black) and match the result against the contents of m_3. Begin by dividing m_8 by 2 and taking the fractional part of the result, which will be 0 if m_8 is even and .5 if it is odd. Next multiply by 2, giving a result of 0 if m_8 is even and 1 if it is odd (figures exactly reproducing the odd-even code used for the bet in m_3). Compare this number (0 or 1) with m_3. If the two are equal, the bet wins, and the contents of m_4 should be multiplied by 2. If not, set the contents of m_4 to 0.

The third part of the solution is simply a matter of recalling and displaying the contents of m_8, m_2, and m_4. Note that since the winnings are now in m_2 and m_4, the player can "let the bet ride" on the next spin of the wheel if he wishes.

The display should be set to zero decimal places for this program. It is possible to get the calculator to output a 00 instead of -1 by checking to see if the contents of m_8 are less than 0 just before they are displayed. If they are, set the display to one decimal place and

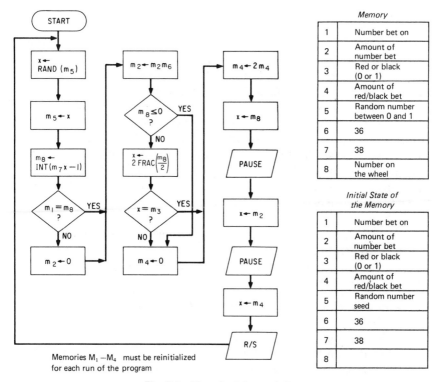

Fig. 3-5 Flowchart for roulette

display a 0. The output will look like this: 0.0. The display will then have to be set back to zero decimal places before displaying the contents of m_2 and m_4. The flowchart is shown in Fig. 3-5. [*Solution was realized on an HP-25 in 43 steps.*]

Note: If you are writing a program to take more than two kinds of bets, you may want to handle your output differently. It would be confusing to have five numbers flash by when the wheel is spun (the number on the wheel and the results of four different bets). You might, therefore, set the calculator to stop after each number. The player can then push R/S to see the next number.

3.7 Blackjack Dealer

Difficulty: 3
Calculator size: Variable

In order to simulate the deal in a game of blackjack realistically, the calculator must store the entire deck of cards and keep track of those dealt. This job is not so difficult as it sounds. In the first place, suits make no difference in blackjack. Second, the face cards (jack, queen, and king) and the 10 all have the same value. Thus, there are only ten *different kinds* of cards: cards worth 1 (ace) through 9 (four of each) and cards worth 10 (16 in all).

This "deck" can be stored in just two memories on your calculator (see Sec. 1.14). Cards 1 through 9 can be coded into a nine-digit number stored in one memory. For instance, the number 1.132243123 can be used to represent a deck with one ace, three 2's, two 3's, two 4's, etc., left in it, whereas a fresh deck would look like this: 1.444444444. The 1 to the left of the decimal point has no "meaning"; it is there solely for a technical reason—to avoid possible errors introduced by the use of the 10^x or y^x key to compute 10^{-k}, the object being to remove a card of denomination k from the deck (see Sec. 1.14). If you have a medium-sized calculator, you may not need this protective device since 10^{-k} can be "built" into a memory that is initialized at 1 and multiplied by .1 each time through the loop being used to search the deck. The number of 10's left in the deck can be stored in the second memory.

Problem: Write a basic program that simulates the deal of cards from a blackjack deck. Each run of the program should return a number from 1 through 10, and the card represented by this number should be removed from the deck stored in the calculator's memory. When the deck is exhausted, the program should return a 0 (or some other appropriate signal).

This is a program that you are not likely to use. The difficulty is that it deals all cards "face up." It can, however, be squeezed onto a small calculator, and it is an interesting exercise. If you have a larger calculator, see if you can improve on the basic program once you have produced it. Perhaps you might assign two user-defined keys to each player. One will deal a card "face down" into a memory and the other "face up" into the display. If you want to be ambitious, design a solitaire game in which the calculator plays the dealer, uses a fixed strategy for hitting itself or not, accepts bets, and keeps track of winnings.

Solution: Instead of attempting to "shuffle" the deck stored in memory, it will be easier to leave it in order and draw cards from it randomly. We do so by generating a random whole number n with a value between 1 and the number of cards in the deck (inclusive), then counting through the deck to the nth card and removing it. Eight memories will be needed, as follows:

m_1 Number of cards from 1 through 9 left in the deck
m_2 Number of 10's left in the deck
m_3 d, the total number of cards left in the deck

m_4 s, a random number "seed"

m_5 Location number of the card drawn in a given run (the number n as defined above)

m_6 Number of cards examined so far in the search for the nth card

m_7 Number of the card being examined in the search for the nth card

m_8 Unexamined portion of the deck (a part of the number stored in m_1)

The program can be broken into three parts: (1) generating the number n, (2) finding the nth card and returning it, and (3) adjusting the contents of the memory registers.

To generate the number n, simply generate a random whole number between 1 and d, without forgetting to change s. (You should know how by now; if not, see Sec. 1.11.) Next, store n in m_5. Decrement d by 1 and initialize m_7 at 1 (ace).

Now to find the nth card. The deck will be "stacked" with the 10's (cards with a value of 10) on top, followed in order by the 1's, 2's, . . . , 9's. Consequently, first recall the number in m_2 and check to see if it is greater than or equal to n. If it is, the nth card in the deck is a 10. Decrement m_2 by 1 and return with a 10 showing. If it is not, store the contents of m_2 in m_6. You must next search the part of the deck stored in m_1. Recall the number stored in m_1, take its fractional part, and store it in m_8. At this point, we pass into the loop that will search the part of the deck containing the cards 1, 2, . . . , 9 (presently stored in m_8). Increment m_7 by 1. On the first pass through the loop, the contents of m_7 will be 1, indicating that the aces are about to be examined. Recall the content of m_8 and multiply it by 10. The effect of the multiplication is to move the decimal point one place to the right and divide the number into two parts. The integer part will be the number of aces left in the deck. Keep the integer part, I, and store the fractional part (the "rest of the deck") in m_8. Add I to the contents of m_6 (keeping the result in m_6), and check to see if the result is greater than or equal to n. If it is, the nth card (or whatever card is now indicated in m_7) is an ace. If not, we have not yet counted through enough cards. Loop back to the sentence, "Increment m_7 by 1." On the second pass through the loop, taking the integer part of $10m_8$ will count the number of 2's in the deck. Proceed through the loop and keep looping back until the content of m_6 plus I exceeds or equals n. When it does, it indicates that the nth card in the deck is the number stored in m_7.

You have now found the nth card in the deck. The only remaining thing to do to remove it from the deck and return with the appropriate number (the content of m_7) showing. If k is the number in

m_7, then this card is removed from the deck by subtracting 10^{-k} from the contents of m_1. This completes the program except for the problem of indicating when the deck has been exhausted, that is, when the content of m_1 is 1.000000000 and the content of m_2 is 0. Go back to the sentence, "Recall the number stored in m_1 and take its fractional part," in the preceding paragraph. At that point, check to see if the fractional part is 0. If it is, the deck is exhausted. (Obviously, the part of the deck stored in m_1 is exhausted, and if the part stored in m_2 were not exhausted, the search for the nth card would have ended before reaching this part of the program.) The flowchart is shown in Fig. 3-6. [*Solution was realized on an HP-25 in 49 steps.*]

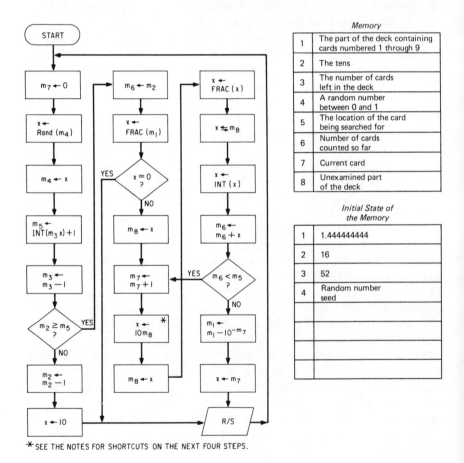

Memory		
1	The part of the deck containing cards numbered 1 through 9
2	The tens
3	The number of cards left in the deck
4	A random number between 0 and 1
5	The location of the card being searched for
6	Number of cards counted so far
7	Current card
8	Unexamined part of the deck

Initial State of the Memory	
1	1.444444444
2	16
3	52
4	Random number seed

* SEE THE NOTES FOR SHORTCUTS ON THE NEXT FOUR STEPS.

Fig. 3-6 Flowchart for blackjack

Notes: Blackjack is one of the few gambling games that gives the player any real chance against the house. (A winning strategy for blackjack has been described by Thorpe in his book, *Beat the Dealer.*)

To be put onto an HP-25 or HP-33E this program has to be squeezed. It is doubtful that it will fit onto a TI-57. Let us point out a few shortcuts made possible by the way things are arranged in Fig. 3-6. Note that m_3—in the fifth box at the left—is decremented by 1 and that there is also a 1 in the box above it. The latter 1 can be used to do the necessary decrementing. Just decrement m_3 with register arithmetic as soon as the 1 is put in the display and before the addition is executed. A good space saving trick in general is to ask yourself if a number put in the display, especially a 0 or 1, can be made to do double duty. A more subtle example of this trick involves the number 10, which happens to appear twice in Fig. 3-6. Ordinarily, putting a 10 in the display takes two program steps since 10 is a two-digit number. Notice, however, that in both of its appearances in Fig. 3-6 the 10 is preceded by a box that places a 1 in the display, and a 10 can be obtained from a 1 in just one step with the 10^x key (INV log x on some calculators).

Finally, look at the five boxes starting with the one marked with an asterisk in the upper right-hand corner. The object of these five steps is to put the fractional part of the number in the display into m_8, retaining the integer part in the display. The sequence given in Fig. 3-6 will do the job, but it is too costly. Here is a sequence that couldn't be described in the flowchart. Note that $INT(X) = X - FRAC(X)$. The trick is to form $INT(X)$ in the display using this formula and to get $FRAC(X)$ into m_8 before the subtraction is executed, as follows:

HP	TI
ENTER	−
FRAC	FRAC
STO 8	STO 8
−	=

For a discussion of this type of trick see Sec. 1.2(10).

3.8   Concentration

Difficulty: 4
Calculator size: Medium

Concentration is a game normally played with a deck of cards that are dealt face down on a table. The players take turns turning up two cards at a time. If the two match, the player who turned them up collects them and gets to turn up two more. If the cards do not form a pair, they are turned back over and the turn passes to the next player. Play proceeds until the deck is exhausted. The player who has collected the most cards wins. Although luck plays a role in concentration, the player with the most accurate memory of cards already turned up almost invariably wins.

A form of concentration can be programmed for the calculator. This version of the game will require that your calculator have indirect addressing (see ' Sec. 1.9). The deck of cards will be replaced by 36 numbers (a sum chosen more or less arbitrarily), two each of the numbers 1 through 18. These will be stored in 36 random memory locations. After the calculator has dealt out the "deck," the players will take their turns by recalling the contents of two of the 36 memories. If the numbers in the memories are the same, the player scores a point and takes another turn.

Since most calculators don't have 36 memories, let alone the number of other memories required for this program, you will need to use multiple storage (see Sec. 1.14). Since each number being stored will have at most two digits, you can pack four such numbers into a single memory. (In fact, since most programmable calculators will accept 10-digit numbers, you could get five 2-digit numbers in each memory. However, another digit, whose purpose is described below, will be required in each memory). The first memory used for multiple storage will then hold the first four numbers. The contents of the memory might be .12071807, for example, thus indicating that the first four numbers are 12, 7, 18 and 7, respectively.

The program for concentration has two basic components. The first is a routine for dealing the 18 pairs of numbers randomly into the 36 memory locations. When you program this part, you will find that the calculator takes a long time to execute it; our first version took 20 minutes. We don't want to give away the solution (skip to the next paragraph if you don't want any hints), but we did discover a trick for speeding things up considerably. Suppose that the calculator is part way through the routine of dealing out the numbers. It goes back for another number and decides to put it in the tenth of 23 remaining empty slots. It would be time-consuming to start at the beginning and search through all the slots until the tenth empty one is reached. To shorten the procedure, attach to each memory register being used for multiple storage a digit to the left of the decimal point that indicates how many empty slots remain in that register. For example, in the middle of the routine, the contents of the first memory might be 2.00070007. The 2 to the left of the decimal point indicates that there are presently two empty slots in that register. This device can greatly facilitate the search for a particular empty slot.

The second component of the program is a routine for recalling the contents of a given slot. When activated by the players after playing begins, it works like this: When a number between 1 and 36, say 19, is entered in the display, it "turns over" the nineteenth card; that is, it goes and finds the number stored in the nineteenth slot and returns with it in the display.

These two routines are enough to enable one to play the game, although the players will have to keep track of the pairs already found and keep score on a piece of paper. The latter operations can be performed by the calculator too, but we recommend that you try to program the fundamental routines first. If you are successful here, you should be able to figure out your own improvements to automate the game to your satisfaction.

Solution: We will first describe the routine needed to set up the game: the random "deal" of 18 pairs of numbers into 36 memory slots. Because four numbers are going to be packed into one memory, nine memories will be needed for the 36 slots. Let these be memories m_{11} through m_{19}. The routine for filling the slots will proceed like this: Numbers 1 through 18 will be generated one after another in a memory and then parceled out, twice each into the 36 slots. A second memory will keep track of how many slots remain to be filled at each stage. This quantity will have two functions. First, when it reaches zero, the routine will be terminated. Second, and more important, it will be used in the procedure for deciding where to put each number.

Suppose that there are 25 slots left to be filled. A little reflection will show that numbers 1 through 5 have already been dealt and that number 6 has been dealt once. We are therefore working on the second 6. When we now generate a random (whole) number between 1 and 25—call it k—number 6 will be dealt into the kth empty slot. The kth empty slot will be found in a two-phase operation. As previously discussed, each memory being used for multiple storage will have a digit to the left of the decimal point indicating how many unfilled slots it has available. Continuing with our example, suppose that k is 7 and that the first four storage memories look as follows:

m_{11}:	2.05000100
m_{12}:	3.00000003
m_{13}:	1.01040006
m_{14}:	2.05000002

By looking at the integer parts of numbers m_{11} through m_{14}, we can determine that the seventh empty slot is in m_{14}. This determination constitutes the first phase of the search for the kth empty slot. The second phase is to go into m_{14}, find the proper empty slot, and put a 6 in it.

Now for the details. The first thing to do is initialize memories m_{11} through m_{19}. Other memories have to be erased, except for a memory containing a random number seed. (Suppose that m_5 contains the random number seed. Then the program will begin by recalling the seed from m_5, clearing all memories, and storing the seed back in m_5.)

Furthermore, memories m_{11} through m_{19} must all be initialized at 4.00000000 (= 4). This operation is accomplished with indirect addressing and a little loop that stores a 4 in m_{11} through m_{19}.

Let's pause here to look at the memories involved:

m_0 Indirect addressing index (HP: I)

m_1 Number of slots left to fill (varies from 36 to 0 during the course of the program)

m_2 Number to be placed in a slot (varies from 1 to 18 during the course of the program)

m_3 k, number of the empty slot into which the number in m_2 is to be placed (a random number between 1 and the number in m_1)

m_4 Empty slot counter

m_5 Random number seed

m_6 Unexamined portion of the fractional part of the contents of the memory m_i where the kth empty slot is known to reside (gets used in the second phase described above)

m_7 $(.01)^j$ where j is 1, 2, 3, or 4 (gets used in the second phase)

m_{11}—m_{19} Storage for pairs of numbers 1 through 18

f_0 Flag indicating whether the number in m_2 has been used once or twice

The program is a sequence of loops. The "big loop" is governed by the contents of m_1, which start at 36 and count down to 0, at which point the routine is terminated. Each time through the big loop a number in m_2 is put into an empty slot indicated in m_3 by means of two smaller loops corresponding to the two phases described above.

Let's assume, then, that the initialization of m_{11} through m_{19} has been done, and it is time to start the big loop. Store the number 36 in m_1. Here we pass into the big loop. Add 1 to m_2 (which will contain a 0 on the first entry into the loop) and raise flag 0, indicating the first of two uses of the number in m_2. Next, generate a random whole number between 1 and the contents of m_1. This will be the "random location" where the number in m_2 is to be placed. Store it in m_3. Let k stand for the number in m_3. We now must find the kth empty slot, and we are ready for the next loop.

Store the number 11 in m_0 and a 0 in m_4. Here we pass into the first inner loop. Recall the contents of the memory whose address is in m_0 [TI: RCL IND 0; HP: RCL(i)]. (When the loop is first entered, doing so will recall the contents of m_{11}.) Take the integer part and add it to the contents of m_4 (keeping the result in m_4). Next we add 1 to the contents of m_0, preparing for the next run through the loop. Meanwhile, m_4 is accumulating the number of empty slots available in memories m_{11} through the m_i whose integer part was just added to m_4. Hence, we recall the contents of m_4 and compare them with those of m_3. If the former is smaller, we haven't found the kth empty slot yet. We must loop back to the third sentence of this paragraph. If $m_4 \geqslant m_3$, we *have* found the kth empty slot; it is somewhere in m_i. The next thing to do is go into m_i and find it, which takes us into the next loop.

The function of the second inner loop is twofold. First, we want to find the kth empty slot, but at this point we have counted past it; the number in m_4 is greater than or equal to k. Thus, we will begin by subtracting from m_4 the number of empty slots in m_i. Then we will go into m_i and start looking for the empty slots one at a time (the mechanics of this will be explained below), adding one to m_4 as each empty slot is found. When m_4 equals m_3, the kth empty slot is found. Second, as the above is going on, we will be building a "multiplier," a power of .01, to be used to put the number in m_2 into the proper slot once it has been found. Suppose, for example, that we want to get the number 6 into the second slot in m_i, which presently looks like this: 2.05000002. We must multiply 6 by .0001 $[=(.01)^2]$ and add the result to m_i. The multiplier, .0001, will get built in m_7 during the course of the loop, and m_7 will be initialized at 1. In the loop, the slots in m_i will be examined one at a time, and as each slot is examined, m_7 will be multiplied by .01. Thus, it will contain the proper multiplier when the proper empty slot is found.

The first thing to do is initialize m_7 at 1. Next we want to recall m_i. (If you go back to the paragraph before last, you will see that the number in m_0 is one larger than the index i that we want. This is because m_0 was incremented by 1 immediately after recalling m_i and before it was discovered that m_i contains the kth empty slot.) We get m_i by subtracting 1 from m_0 (the same 1 used in the previous step can be used here) and then indirectly recalling m_i through m_0 [RCL IND 0 or RCL(i)]. Take the integer part of m_i and subtract it from m_4 (see previous paragraph). Take the fractional part and store it in m_6.

Here we enter the second inner loop. We want to examine m_6 two digits at a time. This is done by chopping off the first two digits of m_6 on each run through the loop. Multiply m_6 by 100 (using memory multiplication), thus moving the decimal point two places to the right. Then take the reciprocal of 100 and multiply it into m_7. Recall m_6 and take its integer part, which will comprise the contents of the first slot in m_6. We want to check to see if the slot is empty, but first subtract the number in the display from m_6 (using memory arithmetic so that the number is retained). The effect of this is to chop the first two digits off the contents of m_6. Now we check to see if the number in the display is 0. If it isn't, the slot is filled. Loop back to the sentence, "Multiply m_6 by 100." If it is 0, the slot is empty.

Next we want to know if it is the right slot. Add 1 to m_4 and compare the new m_4 with m_3. If the two are not equal, we haven't found the kth empty slot yet. Loop back to "Multiply m_6 by 100." If they are equal, we have finally found the kth empty slot. All that remains is to store the number in m_2 in the slot. As previously explained, we do so by multiplying m_2 by m_7 (but don't use memory arithmetic in m_2; we want to use the number in m_2 twice) and adding the result to m_i (using indirect addressing). Finally, subtract 1 from m_i. This has the effect of reducing the integer part of m_i by 1, indicating that there is one less empty slot in m_i.

We have now exited the second little loop. It is time to find out what is happening in the big loop. Decrement m_1 by 1 and check to see if m_1 is down to 0 (using the "decrement skip if zero" key if you have one). If it is, since the routine is finished, stop. If not, we need to loop back for another number to store. But first check the flag f_0 to see whether the number in m_2 was just put into a slot for the first or second time. If the flag is up, the number has only been used once. Since we don't want to generate a new m_2, we lower the flag and loop back to the sentence, "Next generate a random number between 1 and the contents of m_1," four paragraphs back. If the flag is down, we do want a new number and must loop back to the third sentence of that same paragraph. This operation completes the first of the two routines in the program.

Luckily, the second routine is easier to describe and to program than the first. It works as follows: A player enters the number of a slot from 1 to 36 in the display. The program then retrieves the contents of that slot. The routine has two phases. In the first phase, it figures out which of the memories m_{11} through m_{19} contains the desired slot and where the slot is located in that memory. In the second phase, it goes in and retrieves the contents of that slot in a two-part chopping operation that chops off all digits preceding and following the ones we want.

To realize the first phase, we need to do a little old fashioned "quotient and remainder" division.[4] Suppose that we want to retrieve the contents of slot number 23. Since there are four slots in each of the memories m_{11} through m_{19}, the twenty-third slot will be the third slot in memory m_{16}. Notice that 23 divided by 4 is 5 with a remainder of 3, and the slot we want is the third in the sixth $(5 + 1)$ memory. In general, if we want to find the kth slot, we divide k by 4, getting a quotient q and remainder r. Thus, $k = 4q + r$. A little reflection will

---

[4] For a thorough discussion of quotient and remainder division, see Sec. 2.11.

show that the kth slot is then the rth slot in the $(q + 1)^{st}$ memory. [The q memories preceding the $(q + 1)^{st}$ contain a total of 4q slots. The $(4q + r)^{th}$ slot is therefore the rth in the $(q + 1)^{st}$ memory). There is a bit of a problem here if r turns out to be 0, but we will discuss that in the detailed description below. In the second phase, we want to retrieve the rth slot in m_{q+1}.[5]

To continue our example, suppose that we want to retrieve the third slot in memory m_{16}, which might look like this: .10071308. First, multiply this number by 10,000 ($= 100^2 = 100^{r-1}$). The result is 1007.1308. Now chop off the digits to the left of the decimal point by taking the fractional part: .1308. Multiplying by 100 yields 13.08, and chopping off the digits to the right of the decimal point by taking the integer part yields 13, which is the number we want. In general, to retrieve the rth slot in memory m_{q+1}, we multiply by 100^{r-1}, take the fractional part, multiply the result by 100, and take its integer part. Now for the details.

Notice that in the above computation, it is $r - 1$, rather than r, that gets used. Thus, it will be convenient if things can be arranged so that $r - 1$ rather than r appears at the end of our quotient and remainder division. This effect can be accomplished by the simple expedient of subtracting 1 from k before carrying out the division, thereby reducing what is "left over" after the division (that is, the remainder) by 1. More important, this trick takes care of the problem arising when r turns out to be zero.

Let's examine this trick a little more closely. Suppose that we want to retrieve the contents of some slot k and when we divide k by 4 we get a quotient q and remainder 0 (don't worry about how the quotient and remainder are found; it will be explained shortly). Then, $k = 4q + 0 = 4q$. In this case, unlike all others, the slot we are looking for is not the rth slot in m_{q+1}; it is the fourth slot in m_q. (If you don't see this, try an example. Suppose that k = 20 and then figure out what q is and where the kth slot is.) Thus, the number identifying the memory that the kth slot is in (namely q) is 1 less than it is in the other cases. Hence, we wish to arrange things in such a way that q gets reduced by 1 for this special case so that we can always use q + 1 to identify the memory containing the kth slot. This is exactly what will happen if we subtract 1 from k before dividing by 4. Since 4 will "go into" k one fewer time, q will consequently be reduced by 1. (Notice further that the remainder will turn out to be 3, which is exactly what it should be). Our basic algorithm for the first phase of this routine, then, will be to subtract 1 from k, divide the result by 4, and put the

---

[5] Here and in what follows, m_{q+1} refers to the $(q + 1)^{st}$ memory being used to store our numbers. Thus, if q is 5, m_{q+1} would be m_{16} rather than m_6.

answer in quotient and remainder form. The quotient and remainder will then be used in the second phase to retrieve the contents of the kth slot.

Now we need to explain the mechanics of quotient and remainder division. Actually, it is quite simple. To divide the number k by the divisor d (in our case d will be 4), first store k (it will be needed later). Next, divide the k still in the display by d, and take the integer part of the result. This will be the "whole number of times" that d goes into k (that is, it will be q). We store q somewhere so that we don't lose it. To find r, multiply q by d and subtract the result from k. The result will be r, and we store it for later use. We might as well store r where we originally stored k since we no longer need k.

At last, we can describe the whole routine! A player will have entered a number k in the display. We want to retrieve the contents of the kth slot. Subtract 1 from k and divide the result by 4, getting the result in the quotient and remainder form described above. Store q and r somewhere. Next add 16 to q, and store the result in m_0. This will be the address of the $(q + 1)^{st}$ memory storing our numbers. Indirectly recall the contents of this memory through m_0 (RCL IND 0), and multiply the result by 100^r (recall that r is now the $r - 1$ of two paragraphs back). Take the fractional part of the result, and multiply it by 100. Finally, take the integer part of this number, and we are done!

There is one annoying technicality that still needs to be considered. How are we going to compute 100^r in the previous paragraph? Should we use the y^x key, the difficulty is that most calculators employ logarithms in their innards to compute y^x, with the result that the answer may not be precisely correct (this is not true of the HP-67, which uses a different algorithm for y^x when y and x are both positive whole numbers, as they are here). This discrepancy can wreak havoc with our multiplier 100^r. To make the answer precise, we need to round it off to the nearest whole number. Some calculators have functions that will do this; consult your owner's manual. If you don't have this function, don't despair; examine Sec. 1.12. (Also see Sec. 1.10 for a further discussion of the y^n key.) The flowcharts for this program are shown in Figs. 3-7 and 3-8. [*Solution was realized on a TI-58 in 205 steps.*]

3.9   How Random Are These Numbers?

Difficulty: 2

In the preceding programs, you have been using random number generators for various purposes, usually to produce a random whole number between 1 and some given number n. The question

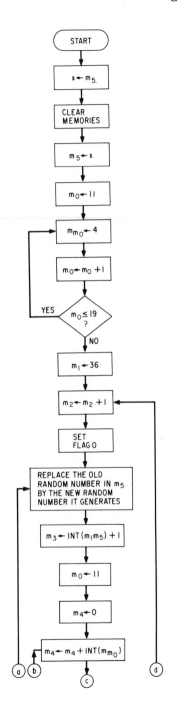

	Memory
0	Indirect
1	# of empty slots
2	# to be placed
3	Slot to be filled
4	Empty slot counter
5	Random #
6	Chopping block
7	$(.01)^j$
8	$k-1$
9	q
10	r
11	
12	
13	
14	
15	Dealt out deck
16	
17	
18	
19	

Fig. 3-7 First flowchart for concentration (cont'd on next page)

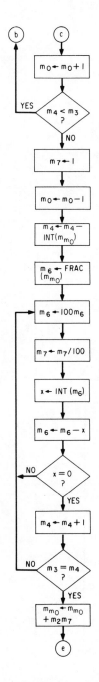

Fig. 3-7 First flowchart for concentration (cont'd)

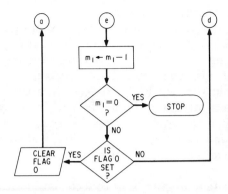

Fig. 3-7 First flowchart for concentration (cont'd)

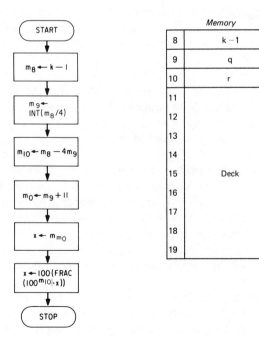

Fig. 3-8 Second flowchart for concentration

arises: Just how random are these numbers? In this section we will describe three tests that you can run.

Suppose that you are generating random numbers between 1 and n and that k is some particular number in this range. The probability of k turning up in any one run of the program should then be 1/n. Thus, in a long sequence of R runs (R being some large whole number), k should turn up R(1/n) times. Since the probability of k turning up twice in succession should be $1/n^2$, in the course of R runs this should happen $(R - 1)(1/n^2)$ times [because k has only $(R - 1)$ chances to turn up twice in a row]. Similarly, k should show up three times in succession $(R - 2)(1/n^3)$ times.

Problem 1:   Write a program that collects data for the above test. It should contain a subroutine that generates random whole numbers between 1 and n, where n is selected by the user (and stored in some memory before the program is run). The user also selects a particular number k in the 1-to-n range and the number of runs R of the random number generator subroutine. When the program is run, it goes through R runs of the above subroutine and counts (1) the number of times that k turns up, (2) the number of double occurrences of k, and (3) the number of triple occurrences of k.

Another problem to consider is how long it will take for your random number generator to start repeating itself. The random number generator starts with a number called the *seed* and applies some process to it to get a new number to which it applies the same process to get a third number and so on. The numbers so generated are both the output of the random number generator and the input used to generate more numbers. Now suppose that after some length of time a number that has already occurred turns up again. Then, since the same algorithm is being applied to generate the next number, the number that followed the original occurrence of this first number will also be repeated on the next run of the algorithm. In fact, the algorithm will repeat the entire sequence of numbers just as it produced them originally. The random number generator will be in a "cycle." In many circumstances, you will want to know the length of the cycle. For example, if you were trying to generate random whole numbers between 1 and 500, you would not want to get caught in a cycle of length 200, because that would mean that at least 300 numbers could not possibly appear.

Of course, any random number generator must eventually repeat itself since there are only finitely many different numbers that it can possibly produce, but the question is, starting from a given seed, how long will it take before some number recurs and how long will the cycle be that starts from the recurrent number. See Sec. 1.16 for a program (the Pollard ρ-method) that solves this problem.

It might seem paradoxical at first, but if your random number generator is truly random, you can get it to generate the number π. Here's how it is done. Use the random number generator to produce a sequence of random points in the "unit square" shown in Fig. 3-9. The unit square consists of points (x, y) both of whose coordinates are between 0 and 1. Some of these points will also fall inside the "unit circle" (which consists of those points whose distance from the origin is $\leq$ 1). The ratio of the number of points falling inside the unit circle to the total number of points produced should be close to the ratio of the area of that part of the unit circle that lies inside the unit square to the area of the unit square. A little calculation will show that the latter ratio is $\pi/4$. Computing the former ratio for a large run of random points in the unit square and multiplying the result by 4 should then give a reasonable approximation of π. The question of how good one should expect the approximation to be will be discussed in the solution.

Problem 2: Write a program for approximating π from your random number generator using the method just described.

Solutions to Problems

Solution to Problem 1: For the data collecting program, you will be given the numbers n, k, and R. These will be put into memories. The complete memory contents will be as follows:

m_1	Random number seed
m_2	Counter for occurrences of k
m_3	Counter for double occurrences of k
m_4	Counter for triple occurrences of k
m_5	Repeat counter (see discussion below)
m_6	k
m_7	n
m_8	R

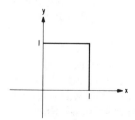

Fig. 3-9 Unit square

The user will initialize m_1, m_6, m_7, and m_8. Memories m_2 through m_5 will start at 0, m_5 being used to indicate whether the last number generated was a k. (More generally, if the number in m_5 was i, then the last i numbers generated are all k's.) This number will be used to produce the count in m_3 and m_4.

Your program needs to do five things: It must (1) generate a random number between 1 and n; (2) check to see if the number is a k; (3) check to see, if it is, whether the last number was also a k; (4) check to see, should the answers to (2) and (3) both be yes, if there has been a triple occurrence of k; and (5) keep track of the number of runs.

Now for the details. First generate a random whole number between 1 and n (we assume that you know how to do this by now). Next, check to see if this number is equal to k (the number in m_6). If it is not, set m_5 to 0 and go to the "end of loop" routine: Decrement m_8 by 1 and check to see if it has gotten down to zero. If it has, the program stops. If not, it goes back to step (1). If the number generated *is* equal to k, add 1 to m_2 and to m_5 (which is counting how many times in a row k has occurred). Now recall the contents of m_5 and check to see if they are ≥ 2; if they are not, got to the "end of loop" routine. If they are, add 1 to m_3 and go on to see if the number in m_5 is ≥ 3. If it is, add 1 to m_4 and then go to the "end of loop" routine. The flowchart is shown in Fig. 3-10. [*Solution was realized on an HP-25 in 39 steps.*]

Solution to Problem 2: Let's think of the problem this way: You make a series of random "shots" into the unit square, and if a shot lands in the unit circle, you count it as a "hit." Tabulate the total number of shots and the total number of hits. The ratio of hits to shots should be approximately $\pi/4$; hence you multiply the ratio by 4 to get an approximation of π. A shot is just a random point in the unit square, which is, in turn, just a pair of random numbers between 0 and 1. The algorithm then, will go as follows: (1) Generate two random numbers between 0 and 1. (2) Take the square root of the sum of the squares of the numbers (this will be the distance of the point from the origin) and check to see if it is less than 1. If it is, score a "hit." Whether or not the shot is a hit, score a shot. (3) (Optional) Calculate the ratio of hits to shots, multiply by 4, and flash the result. (4) Go back to step (1).

There are two possible formats for the program. A program embodying all four parts just described uses the "dramatic effect" format. The user gets to watch as the output slowly converges toward π. In the "research" format, you want to collect data as swiftly as possible, and hence step (3) is skipped. Step (4) can also be augmented by inserting a counter, which is set for a specific number of runs and

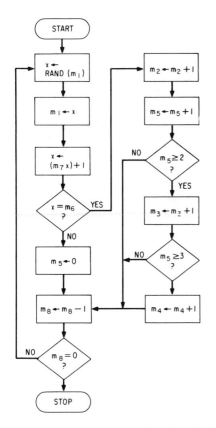

Memory	
1	Random number between 0 and 1
2	Occurrences of k
3	Double occurrences of k
4	Triple occurrences of k
5	Repeat counter
6	k
7	n
8	Loop counter

Initial State of the Memory	
1	Random number seed
2	0
3	0
4	0
5	0
6	k
7	n
8	R

Fig. 3-10 Flowchart for collecting data to test the randomness of numbers

gets decremented after each passage through step (1) and (2), the program being terminated when the count reaches 0. We will give the details of the dramatic effect format, leaving the modifications necessary for the research format to you.

Three memories will be needed:

m_0 random number seed
m_1 number of hits
m_2 number of shots

The program goes as follows: Generate a random number between 0 and 1 and square it. If your calculator has reverse Polish logic, you can let this number sit in the stack until it is ready to be added to the square of the second number. If not, temporarily store the number, say in m_3. Next, generate another random number between 0 and 1, square it, and add it to the previous squared number. Take the square root of the result and compare it with 1. If it is smaller, add 1 to m_1, the "hit" register; if it is larger, skip this step. In either case, go on and add 1 to m_2. Now recall m_1 and m_2 and divide the former by the latter. Multiply the result by 4 and pause with this number in the display. Finally, go back to the beginning of the program for the next run.

Note that the random number generator is used twice in the program. When we programmed the solution, we wrote the random number generator into it twice. This was not very elegant, but, on the other hand, we were not cramped for space. In addition, it saved a little (very little) running time, since subroutines or some sort of flag would be required if the generator were going to do double duty. The flowchart is shown in Fig. 3-11. [*Solution was realized on an HP-25 in 33 steps.*]

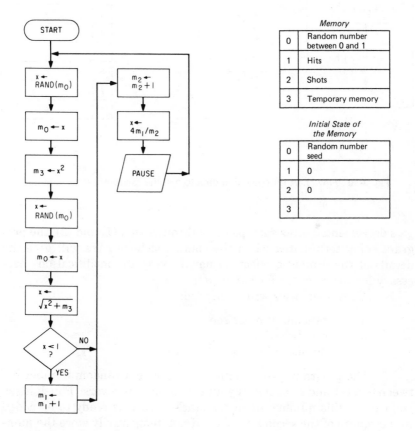

Fig. 3-11 Flowchart for approximating pi from a random number generator

Notes: One step can be saved in the preceding program by modifying the test that determines whether a point (x, y) is in the unit circle. The program as given computes the distance from (x, y) to the origin (0,0), or $\sqrt{x^2 + y^2}$. It then checks whether this distance is less than 1: $\sqrt{x^2 + y^2} < 1$? By squaring both sides of the inequality, one can see that $\sqrt{x^2 + y^2} < 1$ if and only if $x^2 + y^2 < 1$. Thus it will suffice to run the latter test, thereby doing away with the need to take a square root.

If you have run the program for approximating π, you have noticed that it doesn't produce a very good approximation. The obvious question is: How good should one expect the approximation to be if the random number generator is working properly? It is difficult to give a straightforward answer. We know that for longer runs the approximation should improve; but how fast should it improve? Also, if the random number generator is truly random, it conceivably could produce a horrible approximation from an unlucky run. Thus we can say only that it should produce "such and such" an approximation with a certain probability.

It happens, for example, that a run of 100 points should produce an approximation of π with an error no larger than .1 with a probability of .98 (that is, 98 percent of the time the approximation should fall between 3.04 and 3.24). A run of 900 should produce an approximation of the same level of accuracy 99.99 percent of the time. As the desired level of accuracy goes up, however, the number of runs required to have a decent expectation of obtaining that level of accuracy rises dramatically. Table 3-1 summarizes some of these expectations.

Table 3-1 Probability expectations

Length of run	Maximum error	Probability that maximum error will not be exceeded
100	.1	98%
900	.1	99.99%
900	.01	51%
5,000	.01	90%
10,000	.01	98%
10,000	.001	18%
40,000	.001	35%
160,000	.001	65%
1,000,000	.001	98%

Look at the first, middle, and last entries. They are special cases of the following rule: To get an error of less than 10^{-k} with 98 percent certainty, it is necessary to use a run of length 10^{2k}. The last entry says that to get an error less than 10^{-3} with 98 percent certainty, a run of 10^6 is necessary.

Before you set up your calculator to do a run of 1,000,000, consider what will happen if your random number generator starts repeating itself. Since it will then be producing no new information, you cannot expect the level of accuracy to improve from that point on. To determine the maximum degree of accuracy you can expect from your calculator, you will first need to find out how many nonrepeating random points you can produce in the unit square.

3.10   Scorekeepers

Difficulty: 1

Many games require a scorekeeper; this is an ideal job for your calculator. Its memories can keep track of the scores of many players, and it can even keep track of whose turn it is. Not only are these programs useful in themselves, they can also be used as subprograms in other games in this chapter. (We have, for example, included one in our solutions to Basketball and High Low II.)

Since different games need different kinds of scorekeepers, let's start with an easy one.

Problem 1:   Write a program to keep the cumulative score of two players. It should work like this: Suppose that either player can score points at any time. He keys in the points as he scores them. The calculator should then add his points onto his total score and display his new score.

When you try this problem, you may have a little trouble right at the beginning. When the calculator is presented with a score, how does it know whose score it is? (See our solution for the answer.)

In certain games, however, the players take turns on a rotating basis, and each player can score only on his turn. (Scrabble and bowling are examples of such games.) For these games, the trouble just mentioned vanishes. The calculator has only to keep track of whose turn it is; consequently, when a score is entered, it knows whose score it is.

Problem 2:   Write a program to keep score for four players (called 1, 2, 3, and 4 by the calculator) who take turns on a rotating basis. It should work like this. When a player enters the points he has earned at the end of his turn, the calculator should first compute that player's new score and display it for a beat. The program should then stop, displaying either a 1, 2, 3, or 4 to indicate whose turn it is next, and wait patiently for that player's points. If your calculator has user-defined keys, write a routine for one key to review the scores of the four players, one by one.

Problem 3: Here is a variation on Problem 2. Suppose that there are just three players, called 1, 2, and 3 by the calculator. Suppose also that no player's total score ever exceeds 999. Then the entire scoreboard can be displayed at once, in a ten-digit number, as follows:

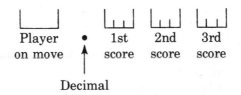

The first digit to the left of the decimal point will be a 1, 2, or 3, depending on whose turn it is. Nine digits to the right of the decimal point are divided into three groups of three, each group containing the score of the corresponding player. For example, if it was the second player's turn, and he had 170 points, while the first player had 21, and the third player 156, the display would show: 2.021170156.

Write a program to keep score in this fashion. As in Problem 2, the three players take turns on a rotating basis, and the program should keep track of whose turn it is.

If you solved Problems 2 and 3 above, your program used the fact that the player's scores were entered in a definite order in order to figure out whose score was being entered. Many games are not like that at all, however, as the next problem shows.

Problem 4: Write a program to keep the cumulative score of six players. It should work like this. Each of the six players has one I.D. number, namely one of the digits 1 through 6. A player can key in what points he has scored at any time as follows. He first keys in his I. D. number and then the points. The program should then sum these points in the appropriate register and halt with the new total score of that player in the display.

Solutions to Problems

Solution to Problem 1: Let us store the first player's cumulative score in m_1 and the second player's score in m_2. When a player enters the points he has just earned, the calculator should add them to either m_1 or m_2, depending on whose points they are. But how is the calculator to know whose points they are? If your calculator has user definable keys, the solution is simple. Each player has his own user-definable key, and hence his own little program to update his score. Each of the two little programs would of course increment the appropriate register (m_1 or m_2) by the number of points keyed into the display and then recall the contents of that register to display the players new score. The flowchart is shown in Fig. 3-12.

If your calculator has no user-definable keys, it has to be told whose points are being entered. A simple way to do this is to key in the first player's points as a positive number and the second player's as a negative number. (A score of zero does not need to be entered!) The routine would begin by testing to see if the points entered were positive or negative. It would then branch to the appropriate one of the two little programs described above. (The second player's points should be *subtracted* from m_2, for they are in the display as a negative number.) The flowchart is shown in Fig. 3-13.

Here is a little frill you can add. Store the player's number (1 or 2) in the first decimal place to the right of the total scores accumulated in m_1 and m_2. Thus, a first player's score of 20 points would be displayed as 20.1; the same total for the second player would be 20.2. For this trick, the registers should be initialized, as follows: $m_1 = .1$ and $m_2 = .2$. To begin the game, this could be done by hand, but it's easier to write a simple routine to do it automatically.

Here is a slightly different version, which will work only if the scores stay below 99999. It allows us to store *both* scores in one register, say m_0. We put the first player's score in the five decimal places to the left of the decimal point and the second player's score in the five places to the right. Thus, if the first player were winning 213 to 190, then $m_0 = 213.00190$. To update m_0, the first player's points are simply added to m_0. But the second player's (negative) points should first be multiplied by 10^{-5} and then subtracted from m_0. [We recom-

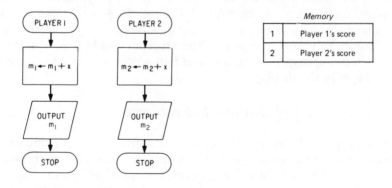

Fig. 3-12 Flowchart for keeping the cumulative score of two players

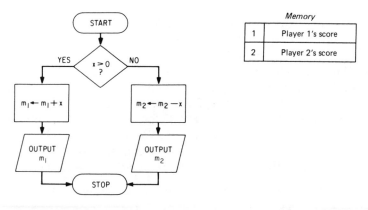

Memory	
1	Player 1's score
2	Player 2's score

Fig. 3-13 Flowchart for program of Fig. 3-12 using calculator without user-definable keys

mend that you initialize this program by storing 10^{-5} (.00001) permanently in register m_3 rather than having the program make the number anew every time; see Sec. 1.2.] The flowchart is shown in Fig. 3-14.

Solution to Problem 2: Here is a simple solution. We use four registers—m_1, m_2, m_3, and m_4—to keep track of the cumulative scores of the four players. And we write four routines, one for each player. Each routine first sums the display into the appropriate register, recalls the new contents of that register for a beat, then places the number of the next player (1, 2, 3, or 4) into the display and halts. These four routines are simply stacked up on top of one another with an appropriate GTO statement at the bottom of the stack to reset the program pointer at

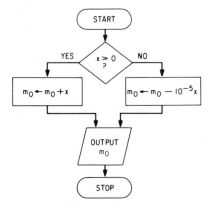

Fig. 3-14 Flowchart for program of Fig. 3-12 allowing both scores to be stored in one register

the top again. Thus, after each player keys in his score, the program halts at the beginning of the next player's routine.

You can add the following frill (as in Problem 1) to this program. Don't just store each player's score but also his number in the first decimal place to the right of the decimal point. Thus, a display of 948.3 would mean that the third player has a total score of 948. (This method can be used on an HP 25 for as many as seven players.)

Here is an alternative solution. We will still use the four registers—m_1, m_2, m_3, and m_4—to store each player's score *and* his number. But this time the contents of the register will be as follows:

m_1 Score and player number of the player whose move it is

m_2 Score and player number of the player next in line

m_3 Score and player number of the player next in line

m_4 Score and player number of the player last in line (the player who just completed his move)

Consider the flowchart in Fig. 3-15, which begins with x, the new points just keyed in by the player on the move. The first box

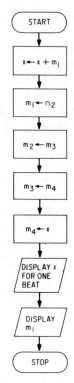

Fig. 3-15   Flowchart for keeping score for four players taking turns on a rotating basis

calculates the player's new score, while the next four boxes "rotate" the contents of the registers. The last two boxes display the new score (and player number) of the player who has just moved and the score and player number whose move it is next, respectively.

Solution to Problem 3:   Let us use register m_0 to store the 10 digit scoreboard, as described in the statement of Problem 3. Our initial task is to add the new points, x, just earned by a player to his total score. What must be added to m_0? Remember that

$$m_0 = \underset{\substack{\text{Player} \\ \text{number}}}{\underline{}} \quad \overset{\uparrow}{\underset{\text{Decimal}}{\bullet}} \quad \underset{\substack{\text{1st} \\ \text{player's} \\ \text{score}}}{\underline{|\,\,|\,\,|\,\,|\,\,|\,\,|}} \quad \underset{\substack{\text{2nd} \\ \text{player's} \\ \text{score}}}{\underline{|\,\,|\,\,|\,\,|\,\,|\,\,|}} \quad \underset{\substack{\text{3rd} \\ \text{player's} \\ \text{score}}}{\underline{|\,\,|\,\,|\,\,|\,\,|\,\,|}}$$

Thus, either $10^{-3}x$ or $10^{-6}x$ or $10^{-9}x$ must be added to m_0, depending on whether it is the first, second, or third player's move, respectively. This player's number is the digit to the right of m_0. The first step, therefore, is

$$m_0 \leftarrow m_0 + (10^{-3})^{\text{INT}(m_0)} (x)$$

Next, we must update the player number. Usually, the step

$$m_0 \leftarrow m_0 + 1$$

will do, except that it changes a 3 to a 4 rather than back to a 1.

Thus, we next test to see if $\text{INT}(m_0) \geq 4$. If it is, we subtract 3 from m_0; otherwise, we leave it alone. In the flowchart shown in Fig. 3-16, we again begin with the number of new points, x, keyed in by the player whose move it is.

Solution to Problem 4:   If your calculator has indirect addressing, the solution is easy. If not, it can still be done, as follows. As in the first solution to Problem 2, each of the four players has his own register $(m_1, m_2, m_3, \text{ or } m_4)$ for his total score (and player number if you like), and each has his own routine in the program to update that register. But again, how does the calculator decide which of the four registers to update? Well, the program will successively subtract 1 from the player I.D. number, testing at each stage to see if it is still nonnegative. If it is, it continues to subtract 1 and test. When the I.D. number finally becomes negative, the program is directed to the appropriate routine.

The flowchart shown in Fig. 3-17 begins with the new points to be accumulated, y, and the I.D. number of the player, x. (If you have an algebraic calculator, y will have to be stored in some register during

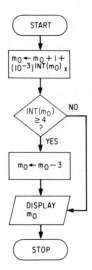

Fig. 3-16 Flowchart for keeping score for three players displaying the entire scoreboard in a ten-digit number

testing.) For technical reasons, this particular program begins by subtracting 2 from the I.D. number. Had we used the test, $x > 0$?, rather than the test, $x \geq 0$?, this would not have been the case. Which version you use depends, of course, on which of the two tests your calculator provides.

3.11   Timers

Difficulty: 2
Calculator Size: Small

Most games requiring some thought (such as chess, checkers, go, backgammon, and scrabble) can be enlivened by the use of a timer. Timers can be of two types; we will consider them one at a time. The simplest one limits the amount of time spent by each player on each move. The players in a chess game may agree for example, to take a maximum of 20 seconds per move. They use a timer to ring a bell 20 seconds after it is reset. A player has to choose his move, make it, and reset the timer for his opponent's move before the bell rings. If the bell ever rings, the player who failed to reset it in time loses, whatever his position on the board.

Your calculator has in its innards a clock that coordinates the myriad tasks the chip must execute. It is not designed for keeping absolute time, only for arranging these tasks in the proper order. Nonetheless, the clock keeps a relatively steady beat, perhaps influenced by such factors as temperature, battery charge, and the age of

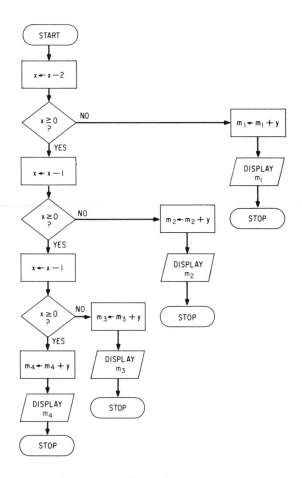

Fig. 3-17 Flowchart for keeping the cumulative score of six players

the calculator. You can therefore expect that if your calculator is called upon to do the same calculation several times, it will take about as long each time.

Problem 1: Write a program that will count down from a positive integer n (input by the user) to 0 and then halt on an error message. It shoud flash consecutively the numbers n, n − 1, n − 2, . . . , 2, 1, and 0 for one beat each and then give an error message. The message signals that a player has overrun his time.

Problem 2: If your solution to Problem 1 was anything like ours, you might have noticed that the time between flashes of n, n − 1, n − 2, . . . , 2, 1, and 0 was relatively constant. Let us call the time between flashes a *unit* of time. As an extra to Problem 1, you might try to extend the unit of time in your solution to some conventional unit, such as 5 seconds. This is easily done. Merely "fatten up" the main loop of your program by making it do a few extraneous calculations. A little experimentation, conducted by timing your program over long runs, should enable you to find a loop that takes close to 5 seconds to execute.

Program your calculator to act as such a timer. The program should work as follows: The players agree on the number n of time units allowed per move and store n in the calculator. The program then counts down from n as described in Problem 1. If left alone, it should reach 0, at which point it will flash an error message to indicate that the player whose move it is has run out of time. Usually, however, a player will stop the clock before this happens, reset it to n, and start it again for the other player. (You should arrange your program so that the fewest possible key strokes are required to stop, reset, and restart the clock.)

The next type of timer does not limit each move but rather the cumulative time a player spends on all his moves. Two clocks, therefore, are needed, one for each player. While a player is pondering his move, his clock is busy counting down and his opponent's clock is still. After he makes his move, he presses a button to shut off his own clock and starts his opponent's clock from where it left off. If a player's clock ever runs out, he loses, whatever the current state of the game. A player begins the game with a large number N of time units, which he is free to apportion among his moves as he likes so long as the total time he has spent thinking on his move does not exceed N units.

Problem 3:   Program your calculator to act as such a timer in a two-player game. At each time unit, it should display all of the following information on one "scorecard" for one beat: (1) The number of time units remaining on the first player's clock should appear in the five decimal places to the left of the decimal point. (2) The number of time units remaining on the second player's clock should appear in the five decimal places to the right of the decimal point. (3) A negative scoreboard indicates that it is the first player's move and his clock is running. (4) A positive scoreboard indicates that the second player's clock is running.

Solutions to Problems

Solution to Problem 1:   The solution is quite easy. Consider the flowchart shown in Fig. 3-18, which begins with $m_0 = n$. The step, $x \leftarrow \sqrt{m_0}$, is the only one requiring any explanation. It is there to produce an error message when the timer has reached $m_0 = -1$.

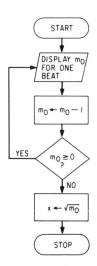

Fig. 3-18 Flowchart for Problem 1 involving timers

Solution to Problem 2: To solve Problem 2, let us suppose that n has been stored initially in register m_1. You need only replace the step in the flowchart of Fig. 3-18 by the sequence shown in Fig. 3-19. Then a player simply stops the program, resets the program counter to the top of the program memory, and starts it off again.

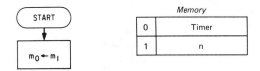

Fig. 3-19 Flowchart for Problem 2 involving timers

Solution to Problem 3: We will use a single register, m_3, to store the number of time units remaining on both clocks, as shown below:

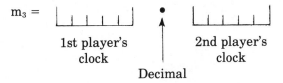

On each run, the main loop of the program will subtract either 1 (to decrement the first player's clock) or 10^{-5} (to decrement the second player's clock) from m_3. The quantity to be subtracted will be stored in m_1. The constant 10^{-5} will be permanently stored in m_2. One more register, m_0, will store either -1 or 1, depending on whether m_3 should be displayed as a negative number (when it is the first player's move) or a positive number (when it is the second player's move).

Now consider the flowchart shown in Fig. 3-20, which must be initialized with $m_0 = 1$, $m_2 = 10^{-5}$, and m_3 equal to the initial configuration of the clocks. (If, for example, each player is to be allowed 7500 time units for the game, you would begin with $M_3 = 7500.07500$.) The first three boxes insure that $m_1 \leftarrow 1$ if it is the first player's turn and $m_1 \leftarrow 10^{-5}$ if it is the second player's. The box $\boxed{m_0 \leftarrow -m_0}$ changes the sign of m_0 ($= \pm 1$) so that when a player has completed his move

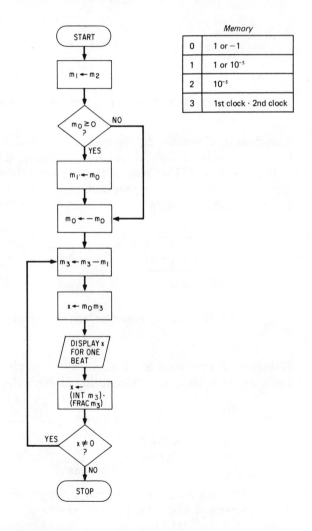

Fig. 3-20 Flowchart for Problem 3 involving timers

and reset the program pointer, the program will automatically switch clocks. The instruction $\boxed{m_3 \leftarrow m_3 - m_1}$ decrements the appropriate clock, whereas the next two boxes display the scoreboard m_3, with the appropriate sign ($m_0 = +1$) for one beat. Note that the next box calculates the product of the two scores, INT m_3 and FRAC m_3. Thus the test

is answered *no* only when one of the two clocks has run out of time.

To use this program, first initialize the calculator as described above and then start the program for the first player. As a player completes his move, he must do three things: (1) stop the program by punching R/S, (2) reset the program pointer to the top of the flowchart, and (3) restart the calculator. If your calculator has user-definable keys, you should assign one of them to this program. Then each player will need to punch only two keys to change clocks, an R/S to stop the program, and then the user-defined key.

3.12   Hockey

Difficulty: 3
Calculator size: Medium (or small)

This is a "simulation" game similar in principle to many games found in bars and electronic game arcades. The game is set up for playing in a plane with x and y coordinates, the "puck" is a point whose position is given by a pair of coordinates, and the "goal" is a segment of the x-axis running between $-r_g$ and r_g (see Fig. 3-21). The "radius" of the goal, r_g, can be set by the players to make the game easier or more difficult. The game has two players, a "shooter" and a "defender." The shooter makes a shot by selecting a pair of coordinates (x_0, y_0) representing the initial position of the puck and the angle α of the shot toward the goal. The defender, on the other hand, has a "block" with which he defends the goal. The block is a movable segment of the x-axis. He positions the block by selecting the x-coordinate of its center; it then extends from $(b - r_b)$ to $(b + r_b)$ along the x-axis (where r_b is the radius of the block; see Fig. 3-21). The radius of the block can also be set by the players.

Once the shooter has made his shot and the defender has positioned his block, the puck is "fired." The program outputs a series of six positions of the puck as it travels towards the goal (the number of positions is arbitrary). The first position is (x_0, y_0), and the last position is $(x, 0)$, when the puck strikes the x-axis. Now, let the x-coor-

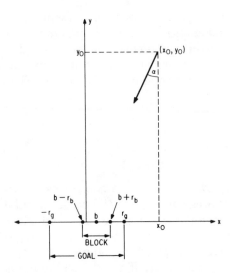

Fig. 3-21   Positioning of the block in hockey

dinate of this point be labeled x_g. If x_g is between $-r$ and r (the easy way to check this is to see if $|x_g| < r_g$), then the shot is "on goal." If the shot is between $(b - r_b)$ and $(b + r_b)$, that is, $|x_g - b| < r_b$), it has been successfully blocked. If not, it scores a goal. The program checks this situation and outputs -1 for a shot that misses the goal, 0 for a blocked shot, and 1 for a goal scored.

This much of the program can be realized on a fairly small calculator, an HP-29 or TI-57. In fact, with some squeezing it can be put onto an HP-25 (see the Appendix). Set the program up so that it will take the data one number at a time, stopping for each piece of input (see Sec. 1.5). Or, with a reverse Polish calculator, you can load the data into the stack and save memories. If you have a larger calculator with user-defined keys and memories to burn, you can use them to take and store the data. This approach has the advantage that if you want to change only one or two numbers on your next shot, you won't have to key in all the data over again. The extra space of a large calculator also allows you to set up a pair of scorecards, one for each player. Player 1's scorecard will work like this: If he has just scored a goal, there will be a 1 in the display. The scorecard program will multiply the 1 by .01 and add it to the rest of the player's score, returning with a number of the form 1.xx (where xx is his cumulative score). Player 2's scorecard will work similarly, returning with a number of the form 2.yy.

Solution: Our solution falls into two phases: (1) collecting and processing the data, and (2) outputting successive positions of the puck and the result of the shot $(-1, 0, \text{ or } 1)$. Two memories will be used to store the successive positions of the puck, which will be equally spaced so that the distance between consecutive positions will be one-fifth of the total distance traveled by the puck between (x_0, y_0) and $(x_g, 0)$. The distance is one-fifth rather than one-sixth because there are five intervals between the six positions of the puck).

Call the distance between successive x-coordinates of the puck Δx and between successive y-coordinates Δy. From Fig. 3-22 (and a little trigonometry), one can see that $\Delta y = y_0/5$ and $\Delta x = \Delta y (\tan \alpha)$ $= (y_0/5)\tan \alpha$. Both Δy and Δx will be placed in memory. Each pair of coordinates can then be obtained from the preceding pair (x,y) by subtracting Δy from y and Δx from x (the puck is moving in a "negative direction" as it goes toward the goal).

We also need to compute x_g. Looking at the drawing once again, one can see that $x_g = x_0 - y_0 (\tan \alpha)$. Incidentally, if x_0 or x_g turns out to be negative, these formulas remain valid. Now let's set up the memories as follows:

m_0 k, a loop counter for the second phase

m_1 x, the x-coordinate of the successive positions of the puck

m_2 y, the y-coordinate of the successive positions of the puck

m_3 Δx, the difference between the x-coordinates of successive positions of the puck

m_4 Δy, the difference between the y-coordinates of successive positions of the puck

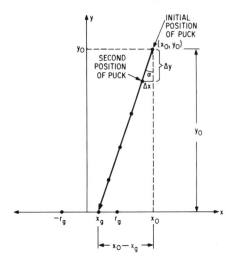

Fig. 3-22 Successive positions of the puck in hockey

m_5 x_g, the x-coordinate of the puck when it strikes the x-axis

m_6 r_g, the radius of the goal

m_7 r_b, the radius of the block

m_8 the result of the shot (-1, 0, or 1)

Both m_6 and m_7 will be set by the users. It is the job of the first phase of the program to initialize the other memories from x_0, y_0, α, and b (which will be keyed in by the players as the program proceeds). We are now ready to describe this phase of the program.

The program starts with x_0 already keyed into the display by the shooter. The first step is to store that number in m_1. Store it also in m_5 [m_5 will eventually contain $x_0 - y_0$ (tan α), which quantity gets built in m_5 as the data comes in]. Next, put in a stop order for the shooter to key in y_0. The next step is to store y_0 in m_2. Divide by 5 and store the result in m_3 and m_4. Stop again to let the shooter key in α. Take tan α, and, using memory arithmetic, multiply it times the contents of m_3, making $m_3 = (y_0/5)$tan α and leaving tan α still in the display. Recall m_2, multiply it by tan α, and (again using memory arithmetic) subtract the result from m_5, which now becomes $x_0 - y_0$ (tan α). At this point, m_1 through m_5 have been initialized as they should be.

Now we want to begin determining the result of the shot. We start by storing a 1 in m_8. Since the contents of m_8 will be adjusted or left alone when we discover whether the shot missed the goal or was blocked, recall the contents of m_6 and compare it with the absolute value of the contents of m_5. If the latter is smaller, the shot is on goal. Leave m_8 alone and branch two sentences ahead to the directive, "Finally, store a 6 in m_0." If not, the shot has missed the goal. Store a -1 in m_8. Finally, store a 6 in m_0. Stop the program again for the defender to key in b, the position of the block.

At this stage, we need to determine whether or not the shot has been blocked. Subtract the contents of m_5 from the contents of the display (b), and take the absolute value of the result. Compare this with the contents of m_7. If the latter is smaller, the shot has missed the block; leave m_8 alone and branch to the next paragraph. If not, the shot has been blocked; store a 0 in m_8. This completes the first phase, and the memories are now set for the output phase.

The bulk of the second phase is a loop that outputs successive x- and y-coordinates of the puck as it travels towards the goal. Both m_1 and m_2 are recalled and displayed, then decremented by m_3 and m_4, respectively. At the bottom of the loop, m_0 is decremented by 1. When m_0 reaches 0, the loop is terminated and the program stopped with the contents of m_8 in the display. Following are the details.

First set the display to two decimal places. This is about as much as one can assimilate when the coordinates are flashed. The setting is done within the program because it will get changed below. Now the more interesting of the two coordinates of the puck is the x-coordinate, since the final x-coordinate of the puck determines the result of the shot. The y-coordinate always starts at y_0 and goes down in equal steps to 0. Consequently, it is more or less just a reference mark. This being the case, we prefer to output the y-coordinate first, followed by the x-coordinate. Start, then, by recalling m_2 and pausing with it in the display for one or two beats, depending on how long the PAUSE on your calculator is. Do the same for m_1. Now decrement m_1 and m_2 by the contents of m_3 and m_4, respectively. Decrement m_0 by 1 and check to see if it is down to 0. If not, loop back and display another set of coordinates. If it is, set the display to zero decimal places, recall the contents of m_8, and terminate the program. The flowchart is shown in Fig. 3-23. [*Solution was realized on a TI-58 in 92 steps.*]

Note: If you are trying to do this on an HP-25, you can dispense with memories m_6 and m_7 by keying r_g and r_b directly into the program where they are needed.

3.13   Basketball

Difficulty: 3
Calculator size: Medium

This long program requires a calculator with user-definable keys. We also recommend that you first read or work through the previous game (hockey), for the two are similar, and we use some of the same ideas in our solutions to both. Basketball, however, presents a more complicated problem because the motion takes place in three dimensions. The calculator simulates the flight of a basketball toward the hoop and indicates whether the shot was a hit or a miss.

We set the game up in three-dimensional space with x, y, and z coordinates. The "court" will be the x,y plane and the "basket" will be a circle around the z-axis 10 units above the x,y plane (see Fig. 3-24). The "ball" will be a point that starts at the position $(x_0, y_0, 5)$. Both x_0 and y_0 are input by the player, whereas 5 is the fixed initial height of the ball. The player must also input the initial velocity of the ball, v_0; the horizontal angle of the shot, α; and the vertical angle (angle with x,y plane) of the shot, β. Ignoring the effects of air resistance, the height of the ball off the floor at time t is given by the formula

$$z = 5 + v_{z_0}t - 16t^2$$

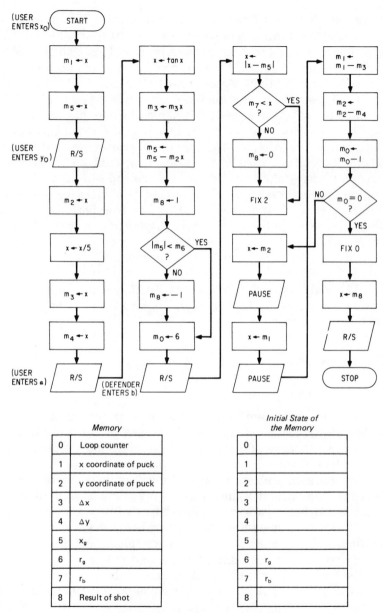

Fig. 3-23 Flowchart for hockey

where v_{z_0} is the vertical component of v_0. The formulas for the x- and y-coordinates of the ball at time t are

$$x = x_0 - v_{x_0}t$$
$$y = y_0 - v_{y_0}t$$

where v_{x_0} and v_{y_0} are the x and y components of v_0. (These formulas
are *differences* rather than sums because the ball moves *toward* the
origin, that is, in a negative direction). The formulas for v_{z_0}, v_{x_0}, and
v_{y_0} are

$$v_{z_0} = v_0 \sin\beta$$
$$v_{x_0} = v_0 \cos\beta \sin\alpha$$
$$v_{y_0} = v_0 \cos\beta \cos\alpha$$

The program will work like this: The player inputs x_0, y_0, v_0,
α, and β, and starts the program. First, it will give six "looks" at the
ball in flight, consisting of six sets of x-, y-, and z-coordinates of the
ball at equally spaced instants of time, starting with the initial position of the ball and ending with its coordinates as it passes through
the x,z plane (the ball should be going through the basket at or near
this time). It will not be necessary to use the formulas for the x- and
y-coordinates. Since the ball will be traveling at constant velocity in
the x- and y-directions, the x and y positions will be equally spaced
between the initial and terminal positions. Thus, the coordinates can
be computed as they were in the previous program (using Δx and Δy).

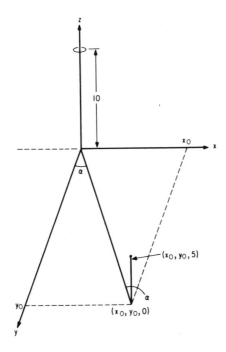

Fig. 3-24 Three-dimensional coordinate system for basketball

However, the program must also figure out whether or not the ball goes through the basket. For this it will be necessary to know the x- and y-coordinates of the ball when it is in the plane of the hoop, that is, when $z = 10$. A little algebra will show that $z = 10$ when $t = (v_{z_0} \pm \sqrt{v_{z_0}^2 - 320})/32$. (Set the value of z at 10 in the formula for z and solve for t using the quadratic formula.) Notice that this gives two values of t—one when the ball passes through the plane, $z = 10$, on its way up, and the other on its way down. The ball should be in the basket at the latter time, when $t_b = (v_{z_0} + \sqrt{v_{z_0}^2 - 320})/32$. If the ball is in the basket at time t_b, the program should output a 2. Otherwise, it returns 0 for a miss.

Finally, there should be "scorecards" for the two players. These will be operated by a pair of user-defined keys, say A and B. If the first player scores two points, he presses A when the 2 is returned by the master program and "collects" his points. The calculator then returns a three-digit number, 1.xx, which indicates that he has xx points. The second player's scorecard is operated by pressing B and returns a similar score, of the form 2.yy.

Solution: We want the main program solving this problem to do three things: (1) initialize the memories, (2) compute and display the path of the ball, and (3) determine whether or not the ball has gone through the basket. Five additional short programs will be used for storing the variables x_0, y_0, α, β, and v_0. For example, key A might operate a program that simply stores whatever is in the display (x_0) in memory register 1. Of course, the player could do the storage by hand, but this way he needn't remember where x_0 is supposed to be stored. Other short programs that will be used for keeping score are described later.

Now let's consider the main program. In order to make its computations, the program will need the following numbers, all of which will be stored in memory:

m_1	x_0, the x-coordinate of the ball's initial position
m_2	y_0, the y-coordinate of the ball's initial position
m_3	α, the horizontal angle of the shot
m_4	β, the vertical angle of the shot
m_5	v_0, the initial velocity (speed) of the ball

These numbers are put in memory by the player before the program is run (using, let's say, keys A, B, C, D, and E). From them, the main program and its subroutines will compute and store the following:

m_6	v_{z_0}, the vertical (z) component of the ball's initial velocity

m_7	v_{x_0}, the x-component of the ball's initial velocity
m_8	v_{y_0}, the y-component of the ball's initial velocity
m_9	Δt, the length of time between two looks at the ball
m_{10}	t, the elapsed time since the ball was shot (later on, this memory will store the time t_b at which the ball passes through the plane of the basket on its way down)
m_{11}	Δx, the amount by which the x-coordinate of the ball changes between two successive looks at the ball
m_{12}	Δy, the amount by which the y-coordinate of the ball changes between two successive looks at the ball
m_{13}	x, the x-coordinate of the ball at various times
m_{14}	y, the y-coordinate of the ball at various times
m_{15}	z, the z-coordinate of the ball at time t
m_{16}	r_b, the radius of the basket
m_{17}	s_1, the first player's score
m_{18}	s_2, the second player's score

Finally, the memory m_0 will be used as a loop counter during the second phase of the program.

The players initialize memories m_{16} through m_{18}. (We have found that 1 is a reasonable value for r_b.) The first phase of the program initializes the remainder, memories m_6 through m_{15}. We have already given the formulas for the contents of m_6 through m_8 in the statement of the problem. The length of time between successive looks at the ball, Δt, is one-fifth of the time it takes for the ball to get from its initial position to the x,z plane, which in turn happens when $y = 0$. Solving the equation, $y = y_0 - v_{y_0}$, for t when $y = 0$, we get $t = y_0/v_{y_0}$. Thus, $\Delta t = y_0/5v_{y_0}$. We now initialize m_{10} at 0, compute m_{11} and m_{12} as in the solution for hockey (see the flowchart in Fig. 3-25), and initialize m_{13}, m_{14}, m_{15} at x_0, y_0, and 5, respectively. The flowchart for the first phase of the program is shown in Fig. 3-25.

Note that we end with a $\boxed{\text{R/S}}$. This allows the player not to have to watch the calculator while initialization is going on. He can then press $\boxed{\text{R/S}}$ to see the path of the ball when he and the calculator are ready.

The second phase of the program is a loop showing the player six positions of the ball as it moves toward the basket. The x- and y-coordinates are computed from the previous coordinates by substracting Δx and Δy, respectively, as in hockey. The z-coordinate is computed by first updating the time (adding Δt to the previous value of t) and then using the new value of t to compute z from the formula, $z = 5 + v_{z_0}t - 16t^2$. After the coordinates have been computed, they are displayed. The flowchart is shown in Fig. 3-26.

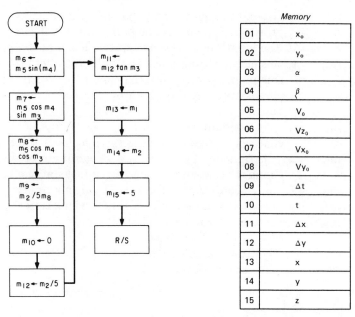

Fig. 3-25 Flowchart for first phase of basketball

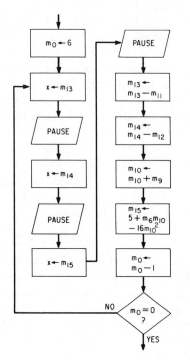

Fig. 3-26 Flowchart for second phase of basketball

The third and final phase of the program determines whether or not the ball went through the basket. This is done by calculating the distance from the ball to the z-axis at the time, t_b, when the ball is coming down through the plane of the basket. The distance should be less than r_b, the radius of the hoop. The distance from a point (x,y,z) to the z-axis is $\sqrt{x^2 + y^2}$. Thus, this phase of the program calculates x and y at time t_b and checks to see if $\sqrt{x^2 + y^2}$ is less than r_b. If it is, the program returns a 2; if not, a 0.

One last detail. Recall that $t_b = (v_{z_0} + \sqrt{v_{z_0}^2 - 320})\,/32$. If $(v_{z_0}^2 - 320)$ happens to be negative, an error message, which may stop execution of the program, will be generated when t_b is computed. To avoid this, the program should check the sign of $(v_{z_0}^2 - 320)$ before extracting its square root. If it is negative, simply branch to that part of the program which outputs a 0. The shot is definitely a miss (what has happened is that the ball has not reached the plane of the basket at all). The flowchart for the third phase is shown in Fig. 3-27.

Last, let's consider the scorecards. The scorecard for the first player will work like this. When the appropriate key is pushed, the program will take the contents of the display (which should be a 2), multiply by .01, and add the result to m_{17}. Then m_{17} is recalled, added to 1, and the result displayed. The display now reads 1.xx, where xx is the first player's cumulative score. If he just wants to see his score, he can enter a 0 in the display and press his scorecard key. The second player's scorecard works similarly. It might also be handy to have a third program that "erases" the scorecards (by setting the contents of m_{17} and m_{18} to 0).

The scorecard programs should also set the display to two decimal places so that the score will look right. In fact, two decimal places make a good setting for the main program, too. If you want to get fancy, you can have the program change the setting from two decimal places when displaying coordinates of the ball (this being about all one can assimilate when the coordinates are flashed) to zero when outputting the 0 or 2. [*Solution was realized on a TI-58 in 00 steps.*]

Notes: The program can be shortened in several places, but we have avoided such subtleties in order to keep our presentation straightforward.

There is a technicality that we have overlooked in this solution. Although it is very unlikely to come up in practice, it is probably worth mentioning. It is possible for the ball to go through the hoop on its upward passage through the plane, $z = 10$, and then drop back through the hoop on the way down. This is not a likely event unless you shoot from very close to the basket. Technically, such a shot should be a miss, but the program would score it as a basket since the only

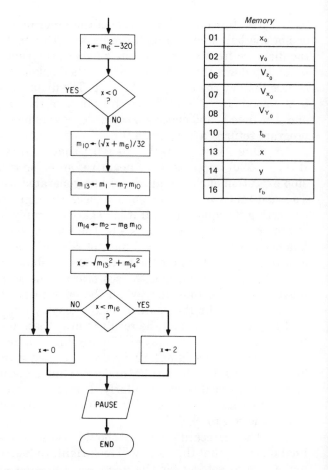

Fig. 3-27   Flowchart for third phase of basketball

thing the program tests is what is happening when the ball is on its way down. The ball passes through the plane of the basket on its way up at time t', where $t' = (v_{z_0} - \sqrt{v_{z_0}^2 - 320})/32$. If you want to write the program to exclude the type of shot just described, insert a test in the third phase that makes sure that the ball is *outside* the hoop at time t' (that is, $\sqrt{x^2 + y^2} > r_b$ at time t').

The formulas for the path of a projectile we have used in this problem were discovered by Galileo (1565–1642), who is regarded by many historians of science as the first modern physical scientist. His account of the motion of projectiles was characteristic of his approach to physical problems in general. He confined his attention to what he could quantify and describe mathematically. He dissected each problem into its components and made idealizing assumptions, for instance, that the motion under study was taking place in a vacuum.

(The fact that there was heated debate in his day about the very possibility of the existence of a vacuum makes his choice of a vacuum as the setting for his laws of motion all the more revolutionary.) Galileo was in his prime during the early decades of the seventeenth century, the century of the scientific revolution. He died in the year that Newton, perhaps the greatest scientist of all time, was born. Newton is said to have remarked, "If I have been able to see further than others, it is because I have stood on the shoulders of giants." Galileo was one of the giants he had in mind.

3.14   Jelly Beans

Difficulty: 4

In most of the games described so far, the calculator has had a passive role, but there are many games in which we can make the calculator one of the players. (HI-LO II in Sec. 3.5 was such a game.) You can teach your calculator not only the rules, but also how to beat your friends and enemies with a mere flick of a chip!

Consider the following game, played by two children in possession of a bag of jelly beans. They agree to take turns eating some of the jelly beans, with the following restriction. At his turn, the player may eat exactly two, exactly five, or exactly seven jelly beans, no more and no less. As soon as a player has no legal move (that is, when only one or no jelly beans are left), he loses.

Here is a sample game between Jim and Sally, who begin with a bag of 15 jelly beans. It's Sally's move first, and she chooses to eat five jelly beans, leaving Jim with 10 to choose from. (Sally could have also eaten two, leaving 13; or eaten seven, leaving 8; but those were her only other legal choices.) Now it's Jim's turn, and he decides to eat two jelly beans, leaving Sally with eight. She then eats seven jelly beans in one gulp and laughs. For now it is Jim's turn, and he has no legal moves because only one bean remains. Thus, he loses, and Sally wins.

In this game (as well as the two that follow), there are two kinds of positions, *winning* positions and *losing* positions. Briefly, a position is defined as either winning or losing according to how the player *about to move* from the position is faring. Position 1 (that is, the position with only one jelly bean left) is a losing position, for example, because the player about to move (like Jim in the game above) is losing. In fact, he has already lost. Similarly, position 0 (no jelly beans left) is a losing position. Position 2 (two jelly beans left), however, is a winning position; the player on the move can eat the last two jelly beans and win because he has left his opponent in losing position 0. This fact illustrates an important rule governing losing and winning positions.

First Rule: If from any given position there is at least one legal move to a losing position, then that first position is a winning one. We have just used this rule to determine that position 2 is a winning one since there is a move (eating two jelly beans) to a losing position (position 0). Referring to the game between Jim and Sally, we see that position 8 is a winning position because eating seven jelly beans leaves only 1, and 1 is a losing position. Another way of thinking of the first rule is as follows: If there is a legal move that puts your opponent in a losing position, then the position you are in must be a winning one.

Let us start a little list of positions to see if we can figure out which are the winning ones and which the losing:

Position	0	1	2	3	4
Lose or win?	L	L	W	W	

We have labeled 3 with W because by the first rule there is a legal move to losing position 1. What about position 4? It allows only one legal move and that is to eat two jelly beans, leaving one's opponent in winning position 2. Thus, 4 must be a losing position:

Position	0	1	2	3	4
Lose or Win?	L	L	W	W	L

This situation illustrates another important rule governing losing and winning positions.

Second Rule: If all legal moves from a certain position lead to winning positions, then that position is a losing one. This is obviously so because if a player has no choice but to put his opponent in a winning position, he must be in a losing one. Positions 5, 6, 7, 8, and 9 are all winning ones by the first rule. In every case, there is at least one move to one of the losing positions, 0, 1, or 4. From position 10, there are only three legal moves: to 8, 5, or 3. But since 8, 5, and 3 are all winning positions, position 10 is a losing one by the second rule. Here is a list of all positions up to 15:

0	1	2	3	4	5	6	7	8	9	10	11	12	13	14	15
L	L	W	W	L	W	W	W	W	W	L	W	W	L	L	W

No wonder Sally won; she went first from the winning 15!

Now, how did we construct this chart? Suppose that, we have already worked out the status of all positions 0, 1, 2, ..., n − 2, n − 1. We can work out the status of position n as follows. We examine the status of all possible positions to which we can move from n. In the jelly bean game, they are positions n − 2, n − 5, and n − 7. If at least one of these positions is a losing one, then n is a winning position by the first rule. (Furthermore, it is a winning move to move to that losing position.) If, however, each position, $n - 2$, $n - 5$, and $n - 7$, is a winning one, then n is a losing position by the second rule (we have no choice but to place our opponent in a winning position).

Of course, there is nothing special about the numbers 2, 5, and 7. This game may be played with any set A of positive whole numbers; all that is important is that the players decide ahead of time what moves are legal. In the game we described above, A = {2, 5, 7}. In the problems below, we restrict ourselves to sets A whose largest element is at most 10.

Problem 1: Write a program that will generate the sequence of L's and W's for any set of A of numbers between 1 and 10.

Problem 2: Write a program to play and win the jelly bean game for any set A of numbers from 1 to 10. The program should accept as input the current position (that is, the number of jelly beans left when it's the calculator's turn) and determine if it is a winning or losing one. If a losing one, it should "fake it" by eating as few jelly beans as possible. If it is a winning position, it should of course make a winning move.

Solutions to Problems

Solution to Problem 1: Let M denote the largest number in A. The key to this whole program is that the method outlined above to determine the status of position n uses the status of only some of the previous M positions. In other words, if we know the status of any M consecutive positions, we can determine the status of the next one. Since by assumption $M \le 10$, we can store the status of the last M positions in one register, m_1 (see Sec. 1.14). We shall store this as a decimal consisting of 1's and 2's, where 1 denotes a losing position and 2 a winning one. Digits closest to the decimal point represent the more recent positions. For example (referring to the list given above for A = {2, 5, 7}), when the calculator is preparing to calculate the status of 16, m_1 would contain the number .2112212222. The first 2 means that 15 is a winning position; the next two 1's mean that 14 and 13 are losing positions, and so on.

The set A is stored in m_7 as a decimal consisting of 0's and 1's, with a 1 in each of the entries corresponding to elements of A. For example, if A = {2, 5, 7}, then m_7 = .0100101, whereas if A = {1, 2, 3, 10}, then m_7 = .1110000001.

Now, the calculator must determine if there are any losing positions it can move to. This step corresponds to matching up m_1 and

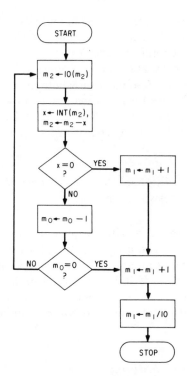

Memory	
0	Counter
1	.wins + losses
2	$m_1 - m_7$
7	.A

Fig. 3-28 Flowchart for generating the sequence of Ls and Ws for any set A of numbers between 1 and 10

m_7 decimal place by decimal place to see if both digits are ever 1. Consider the example of position 16 again when A = {2, 5, 7}. Then, m_1 = .2112212222 and m_7 = .0100101. In the second decimal place of m_7, the digit 1 means that it is legal to move to position 14, and the digit 1 of m_1 means that 14 is a losing position. Thus, since eating two jelly beans at position 16 is a winning move, position 16 is a winning one.

A good way to check a double occurence of 1 is to see if any zeros occur among the first M decimal places of $(m_1 - m_7)$. Consider the flowchart in Fig. 3-28, which begins with $(m_1 - m_7)$ in m_2 and a counter m_0 beginning at m_0 = M. The program "fishes out" the first M digits of m_2 to the right of the decimal point one by one and checks to see if they are zero. If it finds a zero, the current position must be a winning one. If it does not find a 0, the position is a losing one. The flowchart then updates register m_1 accordingly. Let's see how it works.

The first box "pushes" the first digit to the right of the decimal in m_2 out past the decimal point. The next box peels off this digit, calls it x, and then removes it from m_2. Now x is tested to see if x = 0. If it is, we have discovered that the position is a winning one, and we exit from the loop. Otherwise, we decrement the counter with $\boxed{m_0 \leftarrow m_0 - 1}$ and test m_0 to see whether or not we have gone through the loop M times. If not, we go to the beginning of the loop again and fish out the next digit. If we are done, we again exit the loop; this time we know we are in a losing position.

How do we update register m_1 with this new information? Returning once again to our example, if we are determining the status of 16, then m_1 = .2112212222 and m_7 = .0100101. Thus, $m_1 - m_7$, or .2012111222, has a 0 among its first (M = 7) entries, and 16 is therefore a winning position. Consequently, we want to stick a 2 in front of m_1 and "push" everything one place to the right. That is, we want to change m_1 from .2112212222 to .2211221222. (If 16 had been a losing position, we would want to change m_1 to .1211221222 instead.) This is precisely what the following boxes in the flowchart of Fig. 3-28 do:

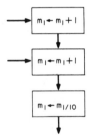

Probably the easiest way to output the sequence of L's and W's is to begin with m_1 = .2222222222 and run the above program ten times, remembering to store M in m_0 and $(m_1 - m_7)$ in m_2 before each run (or set up a loop to have your calculator do this!). Since the first ten L's and W's (that is, 1's and 2's) will now be in the first ten significant figures of m_1, reading from right to left, write them down. Run the program ten more times, and you'll have the next ten 1's and 2's in m_1, and so on.

Solution to Problem 2: To find the status of position n, we must set up a loop to run the solution to Problem 1 for n times. We will use m_3 as a counter for this loop; it will begin with an m_3 of n and count its way down an m_3 of 0. At the end of n runs through the loop, position n will be losing or winning if the first digit to the right of the decimal in m_1 is a 1 or a 2, respectively. Since the operation $INT(10m_1)$ produces this digit, the program tests to see if it is a 1. If it is, the calculator is losing and therefore fakes its position by outputting S, the smallest number in A, which is permanently stored in m_5. Otherwise, the calculator is winning and must now find a winning move.

It goes about doing so as follows: When the program is looking for zeros in $(m_1 - m_7)$, or m_2, it uses m_0 as a counter, counting down from $m_0 = M$ to $m_0 = 0$. When it finds a zero, the number of jelly beans to be eaten by the calculator is equal to the number of times the little loop in Problem 1 has been entered. This number can be recovered from the status of m_0 at the time the digit 0 is found in m_2; in fact, it is $M - m_0 + 1$, and the program should calculate and output this number.

The flowchart in Fig. 3-29 assumes that M, S, and A have been permanently stored in m_4, m_5, and m_7, respectively, and outputs the number of jelly beans the calculator eats when the number n of jelly beans is input. Notice that we began with $m_1 = 1/3$ rather than .2222222222, but it is easier to calculate 1/3, and 1/3 works just as well! (Try it.) [*Solution was realized on an HP-25 in 44 steps.*]

3.15   Whythoff's Game

Difficulty: 3

Here is another game at which your calculator can become an expert and take on all comers. The game begins with two children again in possession of a bag of jelly beans. They have eaten all the good kinds, and the only colors remaining are orange and purple. They decide to take turns eating the rest according to the following rules. A child may on his turn eat either as many orange beans as he likes or as many purple beans as he likes, or he may elect to eat both orange and purple beans. The catch is that if he chooses to eat beans of both colors, he *must* eat the *same* number of each. The child who eats the last bean wins.

Any position in this game can be described by a *pair* of numbers, representing the number of jelly beans of each color remaining in the bag. As in the jelly bean game just described, each position is either a winning or a losing one. (If you have not read the previous section, please turn to it now and read as far as the statement of problem 1. You will need to know what we mean by "winning" and "losing" positions and to understand the first and second rules stated there in order to follow our discussion below.) For example, positions (0,0)[no beans left], (1,2)[one orange bean and two purple beans left], and (5,3)[five orange beans and three purple beans left] are all losing

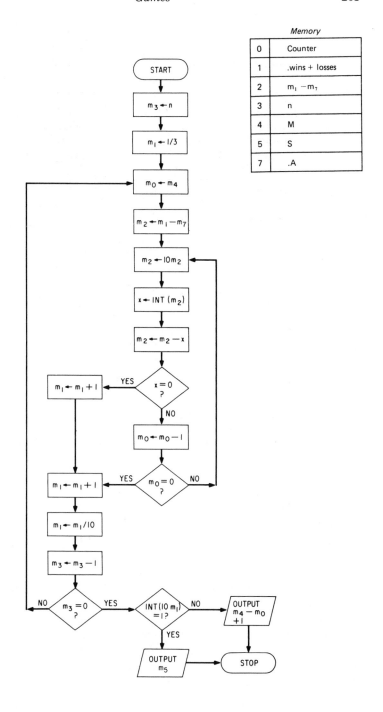

Fig. 3-29 Flowchart for playing and winning the jellybean game

positions, as *you should verify for yourself,* but position (105,103) is a winning position because there is at least one winning move, that is, a move to a losing position. The move is to eat one hundred jelly beans of *each* color (ugh!), leaving your opponent in the losing position (5,3).

There turns out to be a simple way to recognize if position (a,b) is a winning or losing one. (Why the test that we are about to describe works is not an easy question to answer. If you want to find out the reason, read about this game in the book *Ingenuity in Mathematics* by Ross Honsberger. What is important here is that it *does* work, and your calculator can easily apply it.)

Since there is no essential difference between positions (a,b) and (b,a), we can first assume that

$$a \leq b$$

Then (a,b) is a losing position if and only if

$$a = \lfloor \tau(b - a) \rfloor$$

where $\tau = \frac{1}{2}(1 + \sqrt{5})$ and where for any number x, $\lfloor x \rfloor$, or x "round down," is defined to be the greatest integer less than or equal to x. For example, $\lfloor 3 \rfloor = \lfloor 3.1 \rfloor = \lfloor 3.9 \rfloor = 3$.[6] Let's try this rule on (5,3). Here, since a = 3, b = 5, b − a = 2, and $2\tau = 1 + \sqrt{5}$, the result is that $\lfloor \tau(b - a) \rfloor = \lfloor 1 + \sqrt{5} \rfloor = 3 = a$. Hence, (5,3) is a losing position.

You will also need this information: If (a,b) is a losing position with a $\leq$ b, then $a = \lfloor b/\tau \rfloor$ and $b = \lceil a\tau \rceil$ where for any number x, $\lceil x \rceil$, or x "round up," is defined to be the smallest integer greater than or equal to x. (Attention: It may well happen that although $a = \lfloor b/\tau \rfloor$ and $b = \lceil a\tau \rceil$, yet (a,b) is *not* a losing position. For example, $2 = \lfloor 4/\tau \rfloor$ and $4 = \lceil 2\tau \rceil$, but (2,4) is a winning position.)

Problem: Write a program that will enable your calculator to defeat your friends (or enemies) at Whythoff's game. It should work like this: After the player makes his move, he inputs the resulting position into the calculator. The program first tests to see if the position is a losing one. If it is, the calculator "fakes it" by eating one jelly bean of the most frequent color. If the position is a winning one, the program finds a winning move and outputs the resulting (losing) position.

Solution: Our program begins by storing the smaller of the two input numbers, a, in m_1 and the larger one, b, in m_2—using a test to determine which is which. The difference (b − a) is stored in m_3; then $\lfloor \tau(b - a) \rfloor$ is calculated and stored in m_4. (Since we had no room on our HP-25 to calculate τ, this number must be placed in m_0 by the user

---

[5] [x] is calculated using your INT key, as is the function [x] described below. See Sec. 1.12.]

before running the program. If you have a more versatile calculator, you can have it calculate τ here and store it in m_0.)

Next, the program compares a with m_4, or $\lfloor \tau(b - a)\rfloor$. If a = m_4, then the input position is a losing one by the first test described above, and the calculator "fakes it" by outputting (a, b − 1) and halting. Otherwise, the calculator is in a winning position and must decide what sort of a move will leave the opponent in a losing position.

It first decides whether eating a certain number d of jelly beans of each color is a good move. To be a good move, the resulting position (a − d, b − d) must be a losing one. (See the first rule in Sec. 3.14.) By the test described above, (a − d, b — d) is losing if and only if

$$a - d = \tau\lfloor((b - d) - a - d))\rfloor$$

or, equivalently, if

$$d = a - \tau\lfloor(b - a)\rfloor$$

Thus, eating d jelly beans of each color is a good move if $d = a - \tau\lfloor(b - a)\rfloor$ *and* d is a positive number, that is, $a > \tau\lfloor(b - a)\rfloor$. Thus, the calculator tests if $a > \tau\lfloor(b - a)\rfloor = m_4$. If it is, the calculator "eats" $a - \tau\lfloor(b - a)\rfloor$ of each color by outputting the resulting position ($\lfloor \tau(b - a)\rfloor, b - a + \lfloor\tau(b - a)\rfloor$), that is, $(m_4, m_3 + m_4)$ and halting. Otherwise, if $a < m_4$, there are no good moves to be found involving both colors, and the calculator must eat just one color.

Note that the inequality,

$$a < m_4 = \lfloor \tau(b - a)\rfloor$$

only gets worse if a is made smaller. Thus, the calculator must eat from the b color. But how many to eat from b? It follows from the information given in the statement of the problem that the only possible losing positions with one coordinate a are (a, $\lfloor a/\tau\rfloor$) and (a, $\lceil a\tau\rceil$). Thus, the calculator uses the test again, this time to see if (a, $\lfloor a/\tau\rfloor$) is a losing position and outputting it if it is. If (a, $\lfloor a/\tau\rfloor$) is a winning position, the single remaining possibility for a losing position from (a,b) is (a, $\lceil a\tau\rceil$); therefore, the calculator outputs this and halts. How does it know, in this case, that (a, $\lceil a\tau\rceil$) is a losing position? Since (a,b) is a winning one, there must be at least one losing position in which to leave the opponent. Since (a, $\lceil a\tau\rceil$) is the only possibility, it must be the right one!

The flowchart for this solution shown in Fig. 3-30 begins by looking at the two input numbers x and y (a and b). We leave it to you to arrange how the input is entered (see Sec. 1.5).

Notes: The number $(\sqrt{5} + 1)/2$ which occurs in this problem is the famous "golden ratio." Like the even more famous numbers π and e, it has a way of cropping up in unexpected places in mathematics

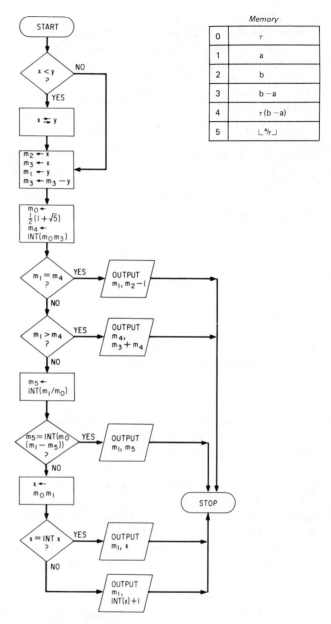

Fig. 3-30 Flowchart for Whythoff's game

and in nature. It is, for instance, related to the "logarithmic spiral," which appears in the shell of the chambered nautilus and the packing pattern of seeds in the sunflower. The ratio was discovered by the Greeks, who defined it as follows.

Let a line AB (see Fig. 3-31) be divided by a point C in such a way that the ratio of the whole to the larger part AC is the same as the ratio of AC to the smaller part CB, that is, AB/AC = AC/CB. The ratio appearing on both sides of the equation is the golden ratio. If we define AB to be of length 1 and let x denote the length of segment AC, then the equation becomes $1/x = x/(1 - x)$ which upon cross multiplication becomes $1 - x = x^2$, or $x^2 + x - 1 = 0$. From the latter equation it readily follows that $x = (\sqrt{5} - 1)/2$ (this number is also sometimes called the golden ratio) and that $1/x$, the golden ratio, is $(\sqrt{5} + 1)/2$ (to see that this last number really is $1/x$, multiply it times x and see what happens).

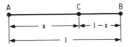

Fig. 3-31 Proportions of the "golden ratio"

Here are two appearances of the golden ratio in mathematics. First, the diagonals of a regular pentagon (see Fig. 3-32) cut one another in the golden ratio. In fact, the pentagram, a regular pentagon with all of the diagonals drawn in, contains scores of occurrences of the golden ratio. And second, the ratios of successive terms of the Fibonacci sequence converge to the golden ratio. You can check this out on your calculator (see Sec. 2.4).

Fig. 3-32 Pentagram illustrating occurrences of the golden ratio

The Greeks were so enamored of the golden ratio that they even worked it into their art and architecture. The rectangle forming the front of the Parthenon is a "golden rectangle." The ratio of its base to its height is the golden ratio. The Greeks thought that this was the most ideally proportioned of all rectangles (and a modern study of the aesthetics of rectangles tends to show that they were right).

3.16   Nim

Difficulty: 4

Not only can you teach your calculator to play nim, you can also teach it how to win! Here are the rules. Two players take turns removing objects from several piles of objects, subject to the following restrictions: A player must remove at least one object per turn, and all the objects removed in a single turn must come from the same pile. When every object has been removed, the next player has no legal move and consequently loses. The interesting thing about nim is that there is a relatively simple strategy for finding the best move every time. We will describe the strategy, and it will be your problem to teach it to your calculator.

Any position in nim can be completely described by a list of nonnegative integers. Each entry in the list represents a pile, and the integer in that entry represents how many objects remain in the pile. For example, the position with three piles left—one pile of five objects and two piles of seven objects each—can be represented by the list (5,7,7). Note that the order of the entries is of no importance; hence, (7,5,7) and (7,7,5) represent the same position. Furthermore, empty piles are of no consequence; hence, (7,5,0,0,7) is no different from our sample position.

We can restate the rules of nim as follows. A legal move consists of choosing a positive entry from the list representing the position of the game and replacing it with a strictly smaller nonnegative integer. If there are no positive components (that is, if the list is (0,0, . . . ,0), then the player loses.

We call a position a *winning* position if the player whose move it is can force a win. For example, (0,0,5,0) is a winning position because he need only remove all five objects from the third pile to leave his opponent with the list (0,0,0,0), thus winning the game. Another example is the list (1,4). We remove three objects from the second pile, leaving (1,1). Since our opponent can take only one object, we take the last one and win.

On the other hand, a position is called a *losing* position if no matter what move we make, we leave our opponent with a winning position. For example, (0,0,0,0) is a losing position, as is (1,1) encountered above. Another example of a losing position is (1,1,1,1); a more complicated example is (1,2,3). Try it! (If you haven't recently read the part of Sec. 3.14 about winning and losing positions, it would be a good idea to do so now.)

When we are faced with a position, we have two tasks. First, we must determine whether it is a winning position or a losing position. Second, if it is a winning position, we must find a winning move—

a move that leaves our opponent with a losing position. Both tasks can be performed with the aid of a bizarre form of addition called *nim addition*. Let us describe how, given two nonnegative integers a and b, we compute a nonnegative integer $a \oplus b$ called the *nim sum* of a and b. We shall proceed by example and compute $21 \oplus 19$.

The first step is to write the summands 21 and 19 in base 2 (see Sec. 1-13): $21_{10} = 10101_2$ and $19_{10} = 10011_2$. Now we "add" up each column separately, using the rules

(1)
$$0 \oplus 0 = 0$$
$$0 \oplus 1 = 1$$
$$1 \oplus 0 = 1$$
$$1 \oplus 1 = 0$$

(You might recognize here the familiar rules

$$\text{Even} + \text{even} = \text{even}$$
$$\text{Even} + \text{odd} = \text{odd}$$
$$\text{Odd} + \text{even} = \text{odd}$$
$$\text{Odd} + \text{odd} = \text{even})$$

Unlike in normal addition, there is *no carry over* from column to column. Thus,

$$
\begin{array}{r}
10101 \\
\oplus\ 10011 \\
\hline
00110
\end{array}
$$

Now, $00110_2 = 110_2 = 6_{10}$; thus, $21 \oplus 19 = 6$.

Your problem eventually, of course, will be to write a program for your calculator to play the optimal strategy in nim.

Problem 1: Write a program that will accept as input any pair of nonnegative integers and output their nim sum, $a \oplus b$.

You should be aware that nim addition shares some of the properties of normal addition. For example,

(2) $$0 \oplus a = a \oplus 0 = a \text{ for any a}.$$

Furthermore, if several integers are to be nim summed, the order is immaterial, that is, for all a, b, and c,

(3) $$(a \oplus b) \oplus c = a \oplus (b \oplus c)$$
(4) $$a \oplus b = b \oplus a$$

In contrast with these familiar properties, nim addition has the following strange property (for any a):

(5) $$a \oplus a = 0$$

In other words, the nim sum of any nonnegative integer with itself is zero!

Now back to the game of nim. Consider the arbitrary position $(a_1, a_2, \ldots, a_n)$. To solve the first task (determining whether a position B is winning or losing) calculate $a_1 \oplus a_2 \oplus \ldots \oplus a_n$. Call this nim sum σ. If $\sigma \neq 0$, the position is a winning one! For the second task (finding the winning move), suppose that the position $(a_1, a_2, \ldots, a_n)$ is a winning one. Then, $\sigma = a_1 \oplus a_2 \oplus \ldots \oplus a_n \neq 0$. We must change one of the a_i's to make the resulting position nim sum to zero. What should we change a_i to if we decide to change it? It turns out that we must change it to $(a_i \oplus \sigma)$, and here's why:

Consider the equation,

$$\sigma = a_1 \oplus a_2 \oplus \ldots \oplus a_i \oplus \ldots \oplus a_n$$

Let us take the nim sum of both sides of this equation with σ:

$$\sigma \oplus \sigma = \sigma \oplus a_1 \oplus a_2 \oplus \ldots \oplus a_i \oplus \ldots \oplus a_n$$

But since by property (5), $\sigma \oplus \sigma = 0$, the equation reads

$$0 = a_1 \oplus a_2 \oplus \ldots \oplus (a_i \oplus \sigma) \oplus \ldots \oplus a_n$$

Therefore, the a_i we decide to change must be changed to $\sigma \oplus a_i$, and the second task consists of finding some a_i so that $\sigma \oplus a_i < a_i$.

Problem 2:   Write a program that will accept as input a nim position. The program should first decide if the position is a winning or losing one. If it is a winning one, the calculator should output the winning move. If it is a losing position, the calculator should not let on but should "fake it" with some innocent move.

Solutions to Problems

Solution to Problem 1:   Because this problem has a difficult solution we we will present it in stages that can be taken as a series of hints.

Hint 1: Let's first recall how to convert a number from base 10 to base 2 (see Sec. 1-13 for another discussion of the process). First divide the number by 2, getting a quotient and a remainder. The remainder (0 or 1) is the rightmost digit in the binary number. Next divide the quotient by 2, getting yet a new quotient and a new remainder. This second remainder (0 or 1) is the second digit from the right in the binary number. This process continues by dividing each last quotient by 2 until the quotient becomes zero.

Let us try the example of converting 55 from base 10 to base 2:

Quotients	Remainders
2	55	1
2	27	1
2	13	1
2	5	0
2	3	1
2	1	1
0	

Consequently, $55_{10} = 110111_2$. (Notice the order in which the 1's and 0's occur.)

Now, how do we go back from base 2 to base 10? We simply write the sequence of powers of 2, beginning 1, 2, 4, 8, 16, . . ., from *right* to *left* above the binary number. Then we add up the powers of 2 that lie above the 1's in the binary number. For example, to convert 110111_2 to base 10, we write the powers of 2 as follows:

32	16	8	4	2	1
1	1	0	1	1	1

and then add to get $32 + 16 + 4 + 2 + 1 = 55$. Not that we have omitted 8 from the sum, because 8 was above a 0 rather then a 1. Thus, $110111_2 = 55_{10}$, as expected.

Hint 2: It seems that there is a lot to be done. There are four basic jobs:

1. Convert a to base 2
2. Convert b to base 2
3. Calculate a $\oplus$ b in base 2
4. Convert a $\oplus$ b back to base 10

Each one of them requires a loop that must be executed several times. Do we need, then, four different loops? We do not; we can use just one loop. Each time through this loop, a small part of each of the four jobs will be done.

Consider the flowchart shown in Fig. 3-33, which is to begin with $m_0 = a$, $m_1 = b$, $m_2 = \frac{1}{2}$, and $m_3 = 0$. We have labeled the steps according to which of the four jobs are being done. There are a few problems with this loop, not the least of which is that it never stops! But let's examine it anyway.

Registers m_0 and m_1 store the successive quotients of a and b, respectively, which are needed in jobs 1 and 2. Thus, we begin by dividing each of them by 2 as described in hint 1. Register m_2 stores the successive powers of 2 needed in job 4; thus we double m_2 each time through this loop. In the next step, y is either 0 or .5 (see Appendix C for a description of FRAC). It is 0 if the remainder on dividing

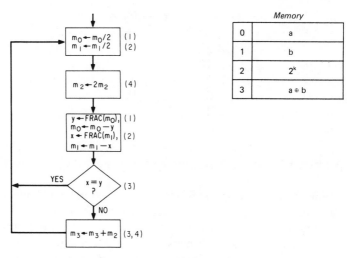

Fig. 3-33   Flowchart for nim loop

m_0 by 2 is 0, and it is .5 if the remainder is 1. The quantity x—also 0 or .5—bears the same relation to m_1, but these remainders are the very numbers we want to calculate in jobs 1 and 2. Referring back to formulas (1), we see that to do job 3, we must compare the remainders. If they are equal, the corresponding binary digits of a and b are equal, and the corresponding binary digit of $a \oplus b$ is 0. If they are unequal, the binary digit of $a \oplus b$ should be 1. We accummulate $a \oplus b$ in m_3. According to the method given in hint 1, to do job 4 we must add a power of 2 to m_3 only when the digit in $a \oplus b$ is 1. This accounts for the test

$$\left\langle \begin{array}{c} x = y \\ ? \end{array} \right\rangle$$

and the box

$$\boxed{m_3 \leftarrow m_3 + m_2}$$

which is only executed if $x \neq y$. See if you can fix this flowchart so that it will stop when its job is done.

   Hint 3: When is the program done? Certainly when both m_0 and m_1 become zero, a condition we could test for every time through the loop. However, we can improve the running time a bit by a simple observation. Consider the following nim sum calculation (in base 2) where b is much larger than a:

$$
\begin{array}{rl}
a = & 1\,0\,1\,1\,0_2 \\
b = & \underline{1\,1\,0\,1\,0\,1\,1\,0\,1\,0_2} \\
a \oplus b = & 1\,1\,0\,1\,0\,0\,1\,1\,0\,0_2
\end{array}
$$

Note that in the leftmost five places (where a has run out of digits) b and a $\oplus$ b agree. Why, then, make the program go through a loop ten times, when on the last five all it's doing is copying the last five digits of b over again in a $\oplus$ b? We can make the calculator exit the loop after the first five digits of a $\oplus$ b have been calculated and then have it attach the rest of b onto the front of these digits. It is a little tricky to figure out how to do this, but here is a hint: The "rest of b" is always being stored in m_1.

Hint 4: The flowchart for our 34-step program on the HP-25 is shown in Fig. 3-34. It begins with numbers a and b and ends with a $\oplus$ b in m_3. The first two steps insure that a $\leq$ b. The next two steps set up the loop described in hint 3. The only difference is the test,

$$\langle m_0 = 0\; ? \rangle$$

allowing the calculator to exit the loop when the binary digits of a have been computed. The step $\boxed{m_3 \leftarrow m_3 + 2m_1m_2}$ turns out to be the way to attach the rest of b onto a $\oplus$ b.

Solution to Problem 2: We will also present the solution to this problem as a series of units.

Hint 1: We suggest that you store the whole position $(a_1, a_2, \ldots, a_n)$ in one storage register as a list to the right of the decimal point. If no pile contains 10 or more objects, one decimal per pile will suffice. If any pile contains 10 or more, but no pile contains 100 or more objects, you will need two decimal places per pile. In the description following, we use ϵ to denote the number of decimal places used per pile. (In the two examples above, ϵ was 1 and 2, respectively.)

Hint 2: Our first job is to calculate the nim sum of all the position numbers. Registers m_0, m_1, m_2, and m_3 will be used to calculate nim sums as described in Problem 1. Register m_5 will begin at zero and accumulate to partial nim sums. Thus, m_5 will successively be storing 0, a_1, $a_1 \oplus a_2$, $a_1 \oplus a_2 \oplus a_3$, etc., until, finally, $m_5 = a_1 \oplus a_2 \oplus \ldots \oplus a_n$. Register m_6 will begin with the list, $m_6 = .a_1a_2a_3 \ldots$ of pile numbers. The a_i's will be successively "fished" out of m_6 and nim summed into m_5, as shown in Fig. 3-35.

The first step begins to fish a_1 out of m_6 by "pushing" a_1 to the left of the decimal point. The next step tests to see if all the a_i's have already been nim summed. If they have, we exit the loop with $a_1 \oplus a_2 \oplus \ldots \oplus a_n$ in m_5. Otherwise, the step $\boxed{m_4 \leftarrow \text{INT}(m_6),\; m_6 \leftarrow m_6 - m_4}$ sets m_4 = current a_i and removes this a_i from m_6. The last step, $\boxed{m_5 \leftarrow m_5 \oplus x}$, uses the program in problem 1. Now try the second (and last) job, to find a good move.

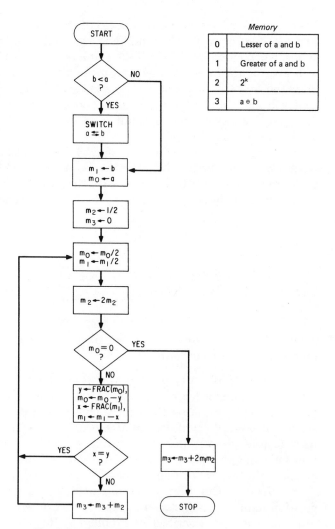

Fig. 3-34   Flowchart for nim sum

Hint 3: The previous job ended with the nim sum, $\sigma = a_1 \oplus a_2 \oplus \ldots \oplus a_n$ in m_5. Assume $\sigma \neq 0$ so that there will be a good move. We must find some a_i so that $\sigma \oplus a_i < a_i$. Figure 3-36 shows how the second job would go, again starting with $.a_1a_2 \ldots$ in m_6. The first two steps are the "fishing" steps, identical with the ones in hint 2. And again we calculate $m_5 \oplus m_4 (= \sigma \oplus a_i)$, but this time we compare it to $m_4 (= a_i)$ to see whether or not $\sigma \oplus a_i \geq a_i$. The program stops with the pile that is to be changed (a_i) stored in m_4 and the number of objects to be left in that pile $(\sigma \oplus a_i)$ stored in m_3. Now how about putting the pieces together?

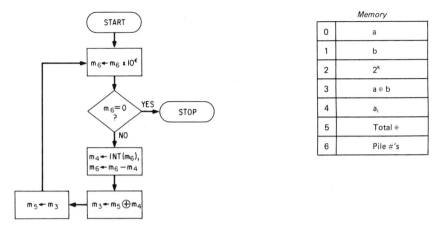

Fig. 3-35 Flowchart for calculating the nim sum of all position numbers

Hint 4: We did manage to fit a program that would perform both tasks into the 49 steps of our HP-25, but we had no room to initialize the registers before each task. Thus, to run this program, the position (and a few other things) had to be keyed in twice, once for each task. Nor did we have room to "fake it" when the calculator was losing. Finally, we could not afford the luxury of leaving the nim sum loop when $m_0 = 0$ and tacking on the rest of b stored in m_1 (as described in hint 3). We had to alter the nim sum program by interchanging m_0 and m_1 so that it would not leave the loop until $m_1 = 0$. This saved six

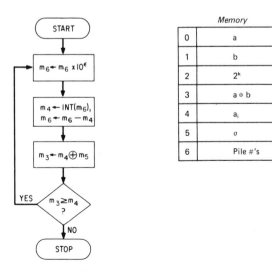

Fig. 3-36 Flowchart for the job in hint 3

program steps at the cost of some running time (the calculation of 2^{10} $\oplus$ 1, for instance, would take 10 runs of the main loop rather than only one. Naturally, if you have a stronger calculator, all these problems are easily avoided.

You have probably noticed that the flowcharts for the two jobs to be done in hints 2 and 3 are almost identical. A great many programming steps can be saved by using some steps twice, once in each job. We will need a flag, however, to tell the calculator which job is being done. We can use m_7. If $m_7 = 1$, then the first job is being done; if $m_7 = -1$, the second job. One more register, m_8, will permanently store the list, $.a_1 a_2 \ldots$.

This program, shown in Fig. 3-37, begins with x, where x =

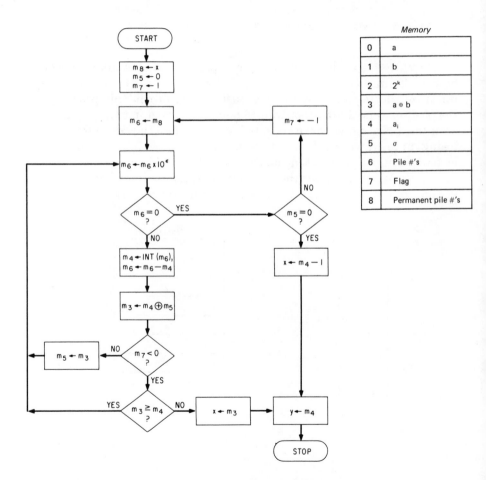

Fig. 3-37 Concluding flowchart for nim

.a_1a_2 . . ., and ends with y, where y = the number of objects originally in the pile from which the calculator has chosen to remove objects and with x, where x = the number of objects *remaining* in that pile after the calculator's move. The first two steps initialize for the first task, and then we enter the loop. Since $m_7 = 1$ at this point, the question,

will always give a "no" answer, and we are simply duplicating the flowchart in hint 2. We exit this loop when $\sigma = a_1 \oplus a_2 \oplus . . . \oplus a_n$ in m_5. We now test to see if $\sigma = 0$. If so, since the calculator is lost, it "fakes it" by removing one object from the last pile. Otherwise, if $\sigma \neq 0$, we set $m_7 = -1$ and enter the loop again for the second task. This time, the question,

$$\langle\ m_7 < 0\ ?\ \rangle$$

is always answered "yes," and we are duplicating the flowchart in Fig. 3-36.

3.17   Number Jotto

Difficulty: 4

There is an old word game called "jotto" or "bulls and cows" that goes like this. One of the two players, the "code maker," secretly chooses any five-letter English word as the "code word." It is the job of the other player, the "code breaker," to guess the code word.

The code breaker picks any five-letter English word as his "guess word." The code maker then gives the code breaker certain information (described below) relating how "close" the code breaker's guess word is to the code word. The more guesses the code breaker makes, the more information he has regarding the code word, and eventually, he will guess the code word. His goal, of course, is to do so with the least number of guesses.

The information the code breaker receives for each of his guesses consists of two numbers, the "bulls" and the "cows." The number of bulls is easy to describe; it is the number of positions in which the code word and the guess word agree. For example, if the code word were ELECT and the guess word EDUCT, the code maker would call three bulls: one because each word begins with E, one because each word has a fourth letter C, and one because each word ends in T. (The code maker does not tell which three of the five positions the three bulls refer to; that's up to the code breaker to figure out!)

As another example, suppose the code breaker next guesses ELATE. He would score two bulls, one for the initial E in both words and one for the second letter L. Now in this case, both the code word and the guess word contain a T. But since the code breaker doesn't have the T in the right place, it does not contribute to the bull count. It does contribute to the cow count, however, for the cows count the number of letters the guess word has in common with the code word, but in the wrong position. Thus, the guess ELATE scores two cows, one for the T's and one for the E's in the third position of ELE*C*T and the fifth position of ELAT*E*. (The E's in the first positions don't count as cows for they have already been counted as bulls.)

In our first example (ELECT and EDUCT), the E in the third position of ELE*C*T does not score a cow with the E of *E*DUCT because this E of *E*DUCT has already been counted as a bull with the first E of *E*LECT. This guess, therefore, does not score any cows.

Since your calculator doesn't talk with letters, but with numbers, let us try a game that we call "number jotto." It's played in the same way as the original game except that we will use an "alphabet" a calculator can understand—the six numbers, 1, 2, 3, 4, 5, and 6—and our "words" will consist of any sequence of four, instead of five, of these six numerical characters, for example, 3264, 5123, 4442, etc.[7]

Problem: Write a program to play number jotto.

The program for this game should do two things: (1) randomly select a four-digit word and hide it in some memory, and (2) accept guesses at the word and return the number of bulls and cows as defined above. Although this game can be realized on an HP-25, it strains the poor thing severely. In the first place, (1) and (2) cannot be put in program memory simultaneously. The program for (1) must be keyed in first and used to generate the code word. Then the program for (2) is keyed in. If you want to play this game on an HP-25, therefore, it would save you a lot of work to have the human opponent fulfill the function of (1). Second, even when (2) has been keyed in, it will require about 25 key strokes for the player to enter each guess. (Of course, if you have a larger calculator, these problems can be avoided.)

Our program will do more than play just four-letter word number jotto; it will also play any number, c ≤ 9, of numbers and words of any length, h ≤ 9. We will stick to c = 6 and h = 4, however, in our description.

---

[7] Master-Mind, a game based on jotto that is marketed by Parker Bros., replaces the letters with colors instead of numbers.

It will be possible for our program to play number jotto with an "alphabet" of not just four, but any number of numerical characters less than or equal to nine, and with "words" of any length from one to nine characters, but we will stick to an "alphabet" of six numbers and a "word" length of four characters.

Solution: If you have tried to solve this problem on your own, you know that it is not easy. Once again, the solution presented here will be in a sequence of steps that can be taken as a series of hints. Read each one carefully in turn and then stop and try to execute that step. Doing so may be enough to trigger the solution in your mind. We should warn you, however, that the steps are not indepndent of one another; they form parts of an integrated whole. If a given step doesn't reveal the complete picture to you, go on to the next step. Even if you just want to understand our solution, you should follow the above procedure in order to see how it all fits together. Our solution was written for an HP-25, which explains why it contains numerous space-saving tricks. If you have a bigger calculator, you will not need these tricks for this particular problem. Nevertheless, they are worth knowing anyway; you might be able to use them later in other programs.

We will begin by describing part (2) of our program, the one that determines the bulls and cows.

Hint 1: Let's not try to do everything at once. We have two main jobs, calculating the number of bulls and calculating the number of cows. Let's write a program to calculate the number of bulls first.

Here are some suggestions. Store a four-character word as a decimal to the *right* of the decimal point, for example, .4113 or .1645. Now fish out one character at a time from the code and guess "words" to see if they are equal.

Hint 2: A flowchart for finding the bulls is shown in Fig. 3-38. Register m_2 count the bulls, beginning at $m_2 = 0$. Registers m_5 and m_6 begin with the code and guess "words," respectively. We will need one more register, m_0; m_0 begins at $m_0 = 4$ and counts down the number of positions to be compared.

The first box, $\boxed{m_0 \leftarrow m_0 - 1}$, simply decreases the counter. The next box,

$$\boxed{\begin{array}{l} m_5 \leftarrow 10m_5 \\ m_6 \leftarrow 10m_6 \end{array}}$$

pushes the leftmost digit of m_5 (the code "word") and m_6 (the guess "word") further to the left, past the decimal point. The next step (after the test

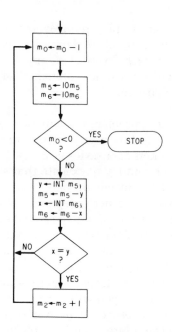

Memory	
0	Counter
2	Bulls
5	.code
6	.guess

Fig. 3-38 Flowchart for finding the bulls

to see if the four positions have been completed yet) removes those digits from m_5 and m_6 and prepares to compare them. If they are unequal, we go back to the beginning and test the next pair of digits. If they are equal, a bull is scored—$\boxed{m_2 \leftarrow m_2 + 1}$—and then the next pair of digits is tested.

Hint 3: Those cows are going to cause a little more trouble. One way to find the number of cows is this: First, we count how many times each digit occurs in the code and the guess "words." For each i, where i = 1, 2, 3, 4, 5, or 6, let λ_i be the number of times i occurs in the code "word" and let μ_i be the number of times i appears in the guess "word." For example, if the code "word" were .1135 and the guess "word" were .5141, then $\lambda_1 = 2$, $\lambda_2 = 0$, $\lambda_3 = 1$, $\lambda_4 = 0$, $\lambda_5 = 1$, and $\lambda_6 = 0$; likewise, $\mu_1 = 2$, $\mu_2 = 0$, $\mu_3 = 0$, $\mu_4 = 1$, $\mu_5 = 1$, and $\mu_6 = 0$.

Now here is the promised formula, Sum (1), which you may have to think about a little before "seeing" it:

$$(1) \qquad \# \text{ bulls} + \# \text{ cows} = \sum_{i=1}^{6} \min(\lambda_i, \mu_i)$$

where $\min(\lambda, \mu)$ denotes the smaller of two numbers, λ and μ. Notice that min (λ_i, μ_i) represents the number of times that the character i occurs in *both* "words." In the example above, although there are no bulls, there are three cows since the sum $\sum_{i=1}^{6} \min (\lambda_i, \mu_i) = \min (2,2) +$ min (0,0) + min (1,0) + min (0,1) + min (1,1) + min (0.0) = 2 + 0 + 0 + 0 + 1 + 0 = 3. Sure enough, the code breaker has guessed three characters right, the two 1's and the 5, but since he didn't put any of them in the right place, they were all cows.

Before we evaluate Sum (1), how about just writing a program to calculate the λ_i's and μ_i's?

Hint 4: We will now consider a program to calculate the λ_i's and μ_i's. It will accumulate the λ_i's in m_3—in the form, $m_3 = 1.\lambda_1\lambda_2\lambda_3\lambda_4\lambda_5\lambda_6$—and the μ_i's in m_4—in the form, $m_4 = 1.\mu_1\mu_2\mu_3 \, \mu_4\mu_5\mu_6$. (The 1 to the left of the decimal is for rounding-off purposes; see the solution of Problem 2 in Sec. 1.14.) We will begin with $m_3 = m_4 = 1$, with $m_0 = 4$, and with m_5 and m_6 containing the code and guess "words," respectively.

In the flowchart of Fig. 3-39, we will explain only the last box since the other steps are identical to the flowchart in Fig. 3-38. (Both flowcharts serve the same purpose, to decrement the counter and fish

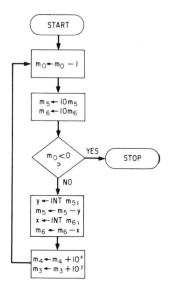

	Memory
0	Counter
3	1. λ's
4	1. μ's
5	.code
6	.guess

Fig. 3-39 Flowchart for calculating the λ's and μ's

out the digits in the code and guess "words" one by one.) This last box,

$$\boxed{\begin{array}{l} m_4 \leftarrow m_4 + 10^x \\ m_3 \leftarrow m_3 + 10^y \end{array}}$$

uses the digits x and y (just fished out from the guess and code "words," respectively) to augment the corresponding μ_i and λ_i. For example, if digit x of the guess "word" is a 3, we must record the presence of this 3 by incrementing μ_3. Thus, we must add $.001(= 10^{-3})$ to $1.\mu_1\mu_2\mu_3\mu_4\mu_5\mu_6$.

The reason we have not written the box above as

$$\boxed{\begin{array}{l} m_4 \leftarrow m_4 + 10^{-x} \\ m_3 \leftarrow m_3 + 10^{-y} \end{array}}$$

is that if we agree to store the code and guess "words" in m_5 and m_6 as *negative* decimals initially, the x's and y's fished out will already be negative and we won't have to change them.

We will, of course, use the 10^x (or y^x) key to execute the above step, but since these keys are often not as accurate as we might like (see Sec. 1.10), we use the trick of storing a 1 to the left of the decimal point in m_3 and m_4, thereby forcing the calculator to round itself off (see the solution to Problem 2 in Sec. 1.14).

Hint 5: Now let's determine how to calculate Sum (1), which was formulated in the hint 3. It actually turns out to be simpler to calculate the quantity,

$$4 - \#\text{bulls} - \#\text{cows}.$$

This is for four character words, of course. With h-character words, we would calculate h $-$ #bulls $-$ #cows.) Notice first that the sums,

$$\lambda_1 + \lambda_2 + \lambda_3 + \lambda_4 + \lambda_5 + \lambda_6$$

and

$$\mu_1 + \mu_2 + \mu_3 + \mu_4 + \mu_5 + \mu_6$$

count the number of digits in the code and guess "words," respectively, color by color. Thus, each sum is equal to 4. From Sum (1), 4

$$- \#\text{bulls} - \#\text{cows} = \sum_{i=1}^{6} \lambda_i - \sum_{i=1}^{6} \min(\lambda_i, \mu_i) = \sum_{i=1}^{6} [\lambda_i - \min(\lambda_i, \mu_i)].$$

Note, however, that $\lambda_i - \min(\lambda_i, \mu_i) = \max(0, \lambda_i - \mu_i)$, where max (a,b) denotes the largest of the two numbers a and b. Therefore,

$$4 - \#\text{bulls} - \#\text{cows} = \sum_{i=1}^{6} \max(0, \lambda_i - \mu_i)$$

Now we will accumulate the above sum in register m_3, as follows. For each i, where i = 1, 2, 3, 4, 5, or 6, we will fish out digits λ_i and μ_i and subtract. If the result, $(\lambda_i - \mu_i)$, is negative, we will go on to the next i. If $(\lambda_i - \mu_i)$ is nonnegative, we add it to m_3 before going on to the next i. Since we must again "take apart" a pair of decimals (this time, the pair—$1.\lambda_1\lambda_2 \ldots$ and $1.\mu_1\mu_2 \ldots$—rather than the code and guess "words"), we suggest storing them in m_5 and m_6, respectively, and again using m_0 as a counter.

Hint 6: Let us first set up the steps shown in Fig. 3-40. All these steps are explained in hint 5 except for the first one. The number of characters, c (here, c = 6), is stored in m_1. The first step stores this value in counter m_0. Since we now have *six* pairs of numbers to check, we begin with $m_0 = 6$.

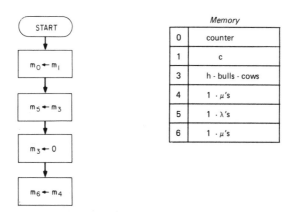

Fig. 3-40 Flowchart for the job in Hint 6

Now let us consider the main part of the program, shown in Fig. 3-41. The first four boxes are identical to some boxes of the flowcharts in hints 2 and 4 and are explained in hint 2. They decrement the counter and fish out the next λ_i, μ_i. The next box calculates $(\lambda_i - \mu_i)$, and the one following that tests to see if $\lambda_i - \mu_i \geq 0$, incrementing m_3 by $(\lambda_i - \mu_i)$ only if $\lambda_i - \mu_i \geq 0$. It then returns for the next i.

With this hint we have reached the point of being able to prepare a final program.

Hint 7: Notice that the three flowcharts in Figs. 3-38, 3-39, and 3-41 have much in common. The first two duties, described in hints 2 and 4, can be done simultaneously. The flowchart in hint 6 (Fig. 3-41) has a block of steps identical to a block of steps in the other

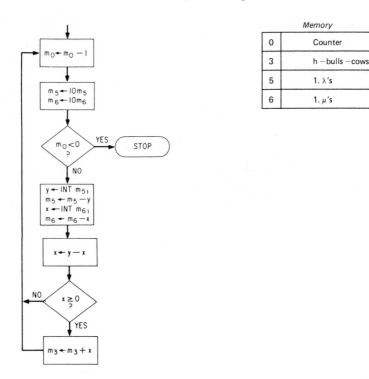

Memory	
0	Counter
3	h −bulls −cows
5	1. λ's
6	1. μ's

Fig. 3-41 Flowchart for the main part of the jotto

two. Thus, we need only put those steps in the final program once although we will use them in two phases.

During the first phase, we will calculate the bulls, $1.\lambda_1\lambda_2 \ldots$, and $1.\mu_1\mu_2 \ldots$, using hints 2 and 4. During the second phase, we will use many of the same steps to calculate the quantity, $4 - \#\text{bulls} - \#\text{cows}$, as in hint 6. We will thus need a flag to tell the calculator what phase it is in. To save a register, we will use m_1 (already being used to store c) as a flag. We will store c in m_1 during phase 1 and $-c$ in m_1 during phase 2.

The flowchart for our 49-step final program for the HP-25 is shown in Fig. 3-42. The column headed "Start" indicates what the storage registers should contain at the beginning of the program, whereas the column headed "Stop" shows what will be in the registers when the program stops.

Notice first that since m_7 permanently stores the code "word," it does not need to be rekeyed for every new guess. Thus, the program begins by moving it to the "chopping block," m_5. Next, the box

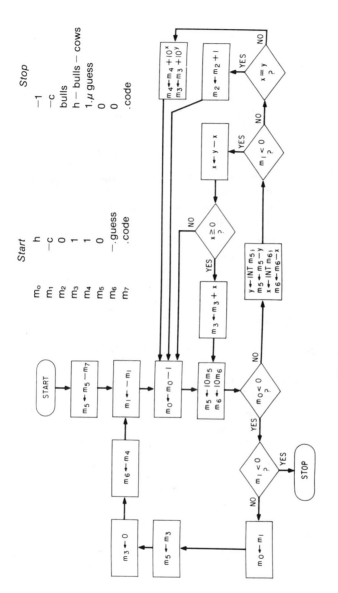

Fig. 3-42 Flowchart for jotto

$\boxed{m_1 \leftarrow m_1}$ places c in m_1, signaling the beginning of phase 1. Notice the presence of

$$\diamond \begin{array}{c} m_1 < 0 \\ ? \end{array} \diamond$$

in two places. These are places where the calculator has to determine what phase it is in.

All other boxes have been previously described, but there is one more thing that must be considered. Let's examine closely how the two jobs described in hints 2 and 4 are combined in phase 1. Note the sequence reproduced in Fig. 3-43. Here, x is a digit of the guess word and y is a digit of the code word. As expected, the box $\boxed{m_2 \leftarrow m_2 + 1}$ registers a bull if x = y, but the $1.\lambda_1\lambda_2 \ldots$ and $1.\mu_1\mu_2 \ldots$ registers are not always incremented, as hint 4 says they should be. They are incremented only when $x \neq y$. Consequently, at the end of phase 1, λ_1 won't count the number of 1's in the code "word" but only the number of 1's in the code "word" that are not counted as a bull. Moreover, $\lambda_2, \lambda_3, \ldots$ and $\mu_1, \mu_2, \mu_3, \ldots$ will be similarly affected.

Nevertheless, it turns out not to matter, for the difference $(\lambda_1 - \mu_1)$ is what we want, and λ_1 and μ_1 will be affected by bulls in the same way. The mistakes will cancel out, therefore, as soon as $(\lambda_1 - \mu_1)$ is calculated in phase 2. The same process will also hold true for the remaining $(\lambda_i - \mu_i)$'s.

The program flowcharted in Fig. 3-42 is as far as we could get on the HP-25. If you have a larger calculator, it should automatically "set up" the registers as shown in the column labeled "Start" on the flowchart. It should also end by recovering the bulls and cows from h, m_2, and m_3. If you have room, you can include a routine to count the number of guesses taken by the user and also to test whether or not the last guess was correct (bulls = h?). If it was, the calculator should indicate this fact and display the number of guesses used.

Picking a random code "word" is easy if you have a few more registers and programming steps. At the beginning, let register $m_7 = 0$. Then choose a random digit between 1 and c by the methods de-

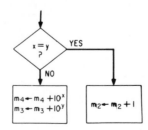

Fig. 3-43 Sequence in Fig. 3-42 where jobs in Hints 2 and 4 are combined

scribed in Sec. 1.11. Add this digit to m_7 and divide the contents of m_7 by 10. Repeat this operation h times, and the result will be a random code word properly stored in m_7.

3.18 Suggestions for Other Games

Dice Probabilities

Difficulty: 2

Write a program that computes the probability of rolling a given number (input by the user) on one roll of a pair of dice. Then write a program that computes the probability of making your point on a given number before "crapping out" in the game of craps.

Note: The latter problem is related to the "problem of points" in gambling. Suppose that a gambling game is interrupted after the stakes are on the table. Player A needs m points to win, and player B needs n points to win. What is an equitable division of the stakes? The solution of this problem in the seventeenth century by Pascal and Fermat led to the mathematical theory of probability.

Water Puzzle

Difficulty: 4

Before attempting this problem, you must first solve the following puzzle. Given an 8-liter bucket and a 5-liter bucket (and an unlimited supply of water), use them to get exactly 6 liters of water. Now write a program that shows how to get z liters of water from an x-liter and a y-liter bucket. You might try to program the problem for an x = 8 and a y = 5 first and then for an x and y that are variable. The program should start with x and y stored in memories. The user enters z, and the calculator displays the successive contents of x- and y-liter buckets until one of them contains z liters. Notice that certain solutions are not possible. For example, if x = 4 and y = 6, it will be impossible to get anything but even numbers of liters from the two buckets. More generally, if d is the greatest common divisor of x and y, then z must be a multiple of d.

Soccer

Difficulty: 3
Calculator size: Medium

The game is similar to hockey (see Sec. 3.12) but is three-dimensional; three coordinates are needed, therefore, to locate the ball in space. The "goal" is a rectangle in the x,z plane. The shooter makes a shot by selecting his position on the field (two coordinates) and direction of shot (two angles). The ball then moves in a straight line toward the goal. The defender tries to block it with a small rectangle that he can place somewhere in the net. If the shot hits the rectangle, it is a blocked shot.

Shooting Gallery

Difficulty: 4
Calculator size: Medium

This is another simulation game. The player will fire a point at a target in the x,z plane. The target consists of a bull's-eye and two outer rings. The player fires from some fixed location, making his shot by selecting a pair of angles (horizontal direction and elevation). Hitting the bull's-eye scores 25 points. Hitting the other two rings scores 10 and 5, respectively. To make the game more interesting, have the calculator select the location of the target randomly; that is, when the appropriate key is pressed, the calculator generates a random pair of coordinates (you may want to fix some bound on the coordinates). The point represented by these coordinates then serves as the center of the bull's-eye.

Tick-Tac-Toe

Difficulty: 4
Calculator size: Large

Write a program for the calculator to play tick-tac-toe. Your first problem will be input and output. Begin by labeling squares 1 through 9 of the playing area to identify them. If you have a printer, you can also get the calculator to do a little graphics. Let 1 represent a move by the calculator, 2 a move by the player, and 0 an empty space. The "board" can then be represented by three three-digit numbers. A partially completed game might look like that shown in Fig. 3-44. The calculator (1) is about to win this game, even if it is the player's (2) move.

$$
\begin{array}{ccc}
1 & 1 & 0 \\
0 & 2 & 0 \\
1 & 2 & 2
\end{array}
$$

The hard part, of course, is to develop a winning strategy that the calculator can be taught. You can simplify things a bit by having the calculator always make the first move. The HP-67 Games Pac has just such a program, in which the first move always begins in the middle of the bottom row (to be at least sporting). Since we have solved the problem of having the calculator make the second move, we don't know how hard it might be (or indeed if it is possible). If you still want to try it, you can simplify things by allowing the player to make his first move only in the middle of the board, in the middle of the bottom row, or in the lower left corner. This structure essentially exhausts all the different moves.

Bowling Scorekeeper

Difficulty: 2
Calculator size: Medium

Write a program that will keep a bowling score for two (or more) players. Each player enters the number of pins knocked down after every ball thrown. The program keeps track of whether it is the player's first or second ball and whether or not there was a spare or a strike. (Note that since the score in a given frame is sometimes not determined until two more balls are rolled by that player, it is not enough to store just the cummulative score of each player!)

Bridge Scorer

Difficulty: 3
Calculator size: Medium

Write a program that will keep a bridge score. It should work like this. At the end of the bidding, the calculator is told who won the contract, what the contract was, and whether or not it was doubled or redoubled. At the end of the hand, the calculator is told how many tricks the declarer won. It then calculates whether or not the contract was made, the penalties for undertricks, and the bonuses for over-tricks, game bids, and slams. The calculator also keeps track of points above and below the line for each side and determines whether or not a side is vulnerable. (Once a side completes a game, the score for that game is accumulated "above the line," and the "below the line" register is cleared for the next game.)

Battleship

Difficulty: 4
Calculator size: Large

In the game *concentration* (see Sec. 3.8), multiple storage was used to store a whole rectangular array of numbers in just a few reg-isters. A similar technique can be used to store the location of the different types of ships employed in the game *battleships*. Either the calculator randomly chooses the placement of its ships, or the human opponent keys in his choices. Then the other player keys in his "bombs," and the calculator computes the hits and misses. (We have not tried this program ourselves, but we are certain that it can be done. The hardest part would probably be the random selection of the locations of the ships.)

Racetrack

Difficulty: 3
Calculator size: Medium

It is not too difficult to simulate a race car in your calculator. You will need six registers for the *racetrack* game: two for the position of the car, two for its velocity, and two for its acceleration. The driver of the car can change its acceleration (only by a limited amount) by keying in the change. The program then uses this information (and Newton's laws of motion) to compute the car's new position, velocity, and acceleration. Several cars can be programmed to race one another on a track. The number of "extras" you can add to the program is limited only by the power of your calculator. Some possible routines you might include are: (1) determining if the player has driven off the track or not, (2) determining whether two cars have crashed, (3) oil slicks on the track (where acceleration immediately becomes zero), (4) lap counters, etc. You might even write a program in which the drivers have to make decisions in "real time"!

Other Racing Games

Difficulty: 3
Calculator size: Medium

The main routines in *racetrack* above—those for calculating position, velocity, and acceleration—can be used to simulate a large variety of racing games. Here are some possiblities.

In *obstacle course,* players have to wind their way through a complicated track without going off the track or hitting one of the many obstacles planted on the track. (The obstacle course would first have to be drawn on paper, and the positions of cars calculated by the program would then be penciled on the paper by the players.)

In *ski slalom,* players have to ski past certain checkpoints (gates) in the right order.

In *bumper cars,* each player tries to hit the vulnerable side of his opponent's car with the invulnerable nose of his own car on an open track.

In *shootout,* both player's are equipped with a gun of a certain range and accuracy, and the two maneuver on an open court shooting at one another.

Learning Engines

Difficulty: 5
Calculator size: Large

If you taught your calculator jelly beans, Whythoff's game, or nim, you probably have found that they are not much fun to play because the silly calculator just keeps winning. To keep it from doing so, you could program it to choose its moves in each game randomly. But this would make *you* a constant winner—a less discouraging outcome, certainly, but just as tiresome after a while.

We have written a program for an HP-67 that strikes a balance between these two extremes. It works like this: The user makes up any two-person game that has 16 or less positions. For example, jelly beans beginning with a bag of 15 jelly beans would do, for there are exactly 16 possible positions (0, 1, 2, . . ., 14, or 15 jelly beans). He then tells the calculator the rules—that is, which positions can legally follow which positions. (The rules can be keyed either one by one or all together from a data card.)

Now the user and the calculator play the game. The calculator begins playing randomly and consequently often loses to the user. It learns from its mistakes, however, and the more it plays, the more it learns. After about twenty quick games, the calculator has become an expert and makes only the best moves every time! Since this is the hardest program we have written to date, we will give you a few clues here.

How does the calculator learn? It simply remembers the last move it made at each turn. Whenever it reaches a position from which there are no legal moves, it is lost and therefore resigns. But just before resigning, it removes this last move (which must have been a *bad* move) from its list of *legal* moves and can therefore never make this particular bad move again.

We number the possible positions in the game 0 through 15. For each position, the calculator must know the list of positions that can legally follow it. This list can be encoded as a five-digit number using the binary storage method described in Problem 3 of Sec. 1.14. Each register can therefore store two such lists—one in the five digits to the right of the decimal point and the other in the five digits to the left.

APPENDIX A

Flowcharts

Flowcharts, a kind of universal programming language originally developed for computers, are also perfectly suited to programmable calculators (which are, after all, tiny computers). A flowchart is not itself a program because it is not written in any particular programming language. Rather, it is a diagram of what the program it represents is supposed to do. Once understood, diagrams of this kind can be readily translated into programs using any language you like. The symbolism of flowcharts is very simple and easily interpreted. That is their virtue.

So let's get right into it. We might as well do the hardest part first, and it will be all down hill after that. Many flowcharts in this book make use of the symbol $\leftarrow$, which means, roughly, "replace by." The name of a memory always appears to the left of the arrow. Whatever appears to the right is to be placed in that memory. For example, $m_1 \leftarrow 1$ means "store a 1 in memory number 1." In a calculator program, this flowchart notation would become the sequence 1, STO 1 (or 1, STO 01). In like manner, $m_1 \leftarrow m_2$ means "store the contents of m_2 in m_1," which would become RCL 2, STO 1 (or RCL 02, STO 01) in a calculator. Note that $m_1 \leftarrow m_2$ says nothing about changing the contents of m_2. After the step $m_1 \leftarrow m_2$ is executed, the contents of m_1 and m_2 will be identical.[1]

Whenever an x appears to the left of an arrow, it always refers to the contents of the display register (we usually don't think of the display register as a memory, but it is). Thus, $x \leftarrow m_1$ means "place the contents of m_1 in the display" (RCL 1 or RCL 01). Since the x-

---

[1] Occasionally, the notation $m_1 \rightleftharpoons m_2$ will be used to indicate that the contents of two memories are to be exchanged.

register is where most calculator computations take place, flowchart steps involving an x often describe a computation—for example, $x \leftarrow m_1 + m_2$ or $x \leftarrow \sqrt{m_1^2 + m_2^2}$. Sometimes an x appears on both sides of the arrow, as in $x \leftarrow \sqrt{x}$ or $x \leftarrow x + 1$. The x to the right of the arrow then refers to the contents of the x-register *before* the step is executed.

A flowchart is drawn as a sequence of different shapes connected by arrows. Each shape represents a program step (or steps). The arrows indicate the order in which the steps are to be executed. The shape

is used to indicate the beginning and end of the program. Most of the shapes in a flowchart, however, are rectangular boxes in which the "normal" steps of the program appear.

A flowchart for a program that computes the areas of circles appears in Fig. A1. Here we assume that the radius of the circle whose area is to be computed will be placed in the display before the program is run. This particular flowchart says nothing about input or output. It only gives a recipe for the central computation that the program has to make: First square the number in the display and then multiply the result by π.

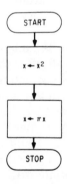

Fig. A1   Flowchart for computing the areas of circles

Flowcharts frequently leave the problem of input and output to the programmer (that's you). However, a special shape has been reserved for input and output steps in a flowchart when they need to be mentioned, a slanted box:

Suppose you wanted to construct the area-of-circles program as a loop. After the area of a given circle has been computed, the program is to stop so that the user can take note of the answer and enter a new radius for the next run. Control will then be sent back to the top of the program for the new run. The flowchart for this procedure might look like that in Fig. A2.

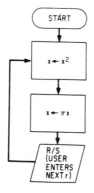

Fig. A2 Flowchart for computing the areas of circles using a loop

It is easy to write a program that will execute this flowchart, as shown below:

HP		TI	
01	x^2	00	x^2
02	π	01	x
03	x	02	π
04	R/S	03	=
05	GTO 01*	04	R/S
		05	GTO 00*

*Your calculator might require a label instead of the "absolute" address given in the GTO order here.

Note how closely the steps in these programs follow the steps in the flowchart.

Flowcharts have great flexibility in the way they allow a computation to be done. For instance, our original flowchart could be rewritten as shown in Fig. A3. Unlike the original, this one does not indicate how πx^2 is to be computed. It merely says, "Compute π times the square of the number in the display and leave the result in the display." The user has to figure out how. In general, the flowcharts

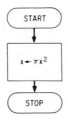

Fig. A3 Rewritten version of Fig. A1

for the easier problems in this book provide more detail than the others.

Next, we need to make some comments about memories. Memories in flowcharts (and elsewhere in the book) are usually denoted by subscripted m's ("memory") and r's ("register"): m_0, r_6, m_7, for example.

Many problems assume that your calculator has eight memories, since all programmable calculators have at least that many. These memories will frequently be labeled m_1, m_2, ..., m_8 or r_1, r_2, ..., r_8. But if your calculator really does have only eight memories, it probably does not have a memory numbered 8. Rather it has a memory at the beginning numbered 0. You will have to watch out for this possibility and change some of our 8's to 0's when you write your own programs.

Second, although it is true that m_6 usually denotes memory number 6 or 06 on your calculator, it does not always do so. In some programs involving lots of memories, we may label one set of memories m_1, m_2, ..., m_k and another n_1, n_2, ..., n_k. Here the subscripts label the numbers that will be placed in these memories, as well as the memories themselves. Thus, it will be up to you to figure out what actual memories on your calculator will be used for m_1, n_1, m_2, n_2, etc. (see Sec. 2.10 for an example).

Third, whenever indirect addressing occurs in a program, we use the notation m_i or m_{m_0}. The former refers to the memory whose subscript is in the indirect addressing register, and m_{m_0} refers to the memory whose subscript is in memory m_0. If m_0 contains a 6, for instance, the m_{m_0} will denote m_6. If m_0 contains a 10, then m_{m_0} will denote m_{10}.

Readers whose calculators have indirect addressing should already know how it works; our point is only to introduce the notation. Others need not worry about it, since they're going to be able to solve the problem anyway (however, see Sec. 1.9 for some tips on how to fake a little indirect addressing on calculators that don't have it).

Now let's get back to flowcharts. Thus far we have introduced three shapes:

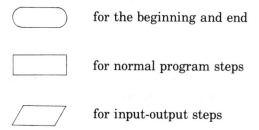

for the beginning and end

for normal program steps

for input-output steps

There is one other special type of step occurring in a program: testing. Test steps are marked by diamonds:

Diamonds have two arrows coming out of them to indicate where control is to be sent next in the program for each possible answer to this test question. Figure A4 is a flowchart of a program to compute the absolute value of a number. (The simpler flowchart in Fig. A5 accomplishes the same thing without a test.)

Since testing works differently in different calculators, you may have to tinker a bit to get your calculator to execute a test (see Sec. 1.3 for a few tips). Most TI calculators have a special test register, the *t-register,* which is used for all testing. To execute the test, $x \geq 0$?, on a TI, the contents of the display will have to be loaded into the t-register first; then the test, $t \geq 0$?, will be executed. HP calculators can execute the test, $x \geq 0$?, directly. Some of the functions of the t-register are performed by the y-register in HP testing.

Occasionally, a flowchart will have a test such as the following: $m_1 = m_2$?. This is one of those "we're leaving it up to you" type steps. No calculator has such a test key; you will just have to figure out how to do it on your particular model.

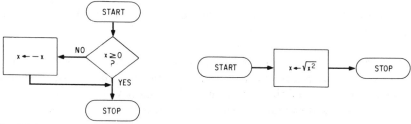

Fig. A4 Flowchart for computing the absolute value of a number

Fig. A5 Simpler version of Fig. A4 without a test

There is one last difficulty with test steps in flowcharts. Because of the differences between calculators, whatever may have been in the x-register before testing began may or may not be there when the test is over. The flowchart, however, makes the (probably unwarranted) assumption that nothing is changed during testing. Consider the following portion of a flowchart shown in Fig. A6. How are you going to execute this test? If you have a TI calculator, you will probably use the sequence, $x \rightleftarrows t$, RCL 1, $x = t$?. For HP calculators, the sequence will be RCL 1, $x = y$?. In both cases, the contents of the x-register will now be the same as the contents of m_1. The *former* contents of the x-register have moved into either the t-register or the y-register. However, the step, $x \leftarrow x + 1$, refers to the former, not the present, x, because the flowchart assumes that the former x would not move during testing. Thus, before executing the step, $x \leftarrow x + 1$, you will have to bring the former x back. We have tried to draw attention to this potential ambiguity whenever it occurs in a flowchart in the text, but you will have to be on the lookout for it, too.

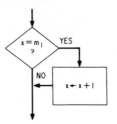

Fig. A6   Portion of a flowchart to illustrate testing problems

It is now well past time to look at some flowcharts close up. In keeping with the philosophy of this book, we will present you with this valuable experience in the form of problems. Solutions appear at the end of this Appendix.

Problem 1:   The input for the program shown in Fig. A7 is a positive number. What does the program do?

Problem 2:   The input for this program will be two positive whole numbers, to be placed in memories m_1 and m_2. The program should divide the larger of the two numbers by the smaller and output the result. Make a flowchart for this program.

Problem 3:   Write a program for your calculator that executes the flowchart made in Problem 2.

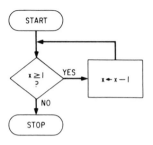

Fig. A7 Flowchart for Problem 1

Most flowcharts in this book are accompanied by *memory diagrams,* tables to the right of the flowchart (usually) indicating the contents of each memory used by the program. These diagrams help you read the flowchart and understand what it is doing. A second memory diagram, labeled "Initial state of the memories," may appear below the main memory diagram to indicate what particular numbers should be in various memories before the program is run. In most cases, the numbers must be placed there by the user. If a memory appears in the main diagram but not in the initial state diagram, then the user need not initialize it since the program will take care of it automatically. Sometimes initialization is indicated within the flowchart itself (see, for example, Fig. 2-17 in Sec. 2.10). The initialization steps that appear in the first box of the flowchart will probably not be actual steps when the program is put on your calculator.

You should be getting the idea by now that flowcharting is a little loose—the price that has to be paid for the flexibility of the language. In most cases, flowcharts can be turned directly into programs for your calculator, but you should always read the accompanying text carefully to make sure that you understand them before writing your programs. Watch out, in particular, for input-output and testing steps. It is here that flowcharts are at their most ambiguous, and, consequently, you must be especially careful. Now let us consider one final problem.

Problem 4: The input for the program shown in Fig. A8 will be two positive whole numbers—a and b. What does the program do?

Solutions to Flowchart Problems

Problem 1: Although the program computes the fractional part of x, we do not recommend using it. It is outrageously inefficient if x is large.

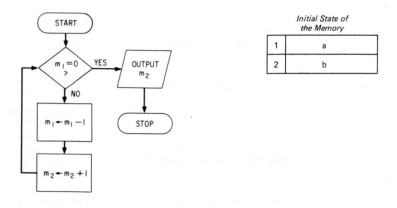

Initial State of the Memory	
1	a
2	b

Fig. A8   Flowchart for Problem 4

Problem 2: Let's call the numbers in m_1 and m_2 "a" and "b," respectively. The straightforward way to solve the problem is to test first to see which number is larger and then branch to the appropriate division. The flowchart is shown in Fig. A9.

Another possible solution is to perform the division first and then check to see whether you did the right thing. This flowchart is shown in Fig. A10.

The program begins by computing m_1/m_2. If this was the proper division, the answer should be greater than 1. If not, the correct answer is m_2/m_1, simply the reciprocal of m_1/m_2. This flowchart contains just as many boxes as the previous one, but you will find that it programs in fewer steps and runs faster. Since the box, $x \leftarrow 1/x$, is a one-step operation, Sec. 1.2(6) can be used to advantage here.

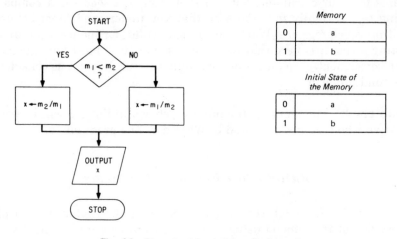

Memory	
0	a
1	b

Initial State of the Memory	
0	a
1	b

Fig. A9   Flowchart for solving Problem 2

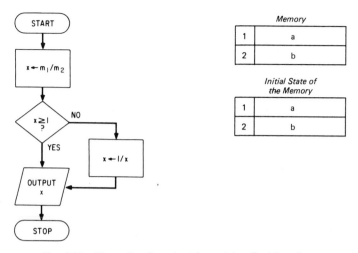

Fig. A10 Alternative flowchart for solving Problem 2

Note: The trick of making a computation, checking to see if you did the right thing, and then adjusting the result if necessary is often times more efficient than the usual sequence of first checking to see which computation should be made and then making it.

Problem 3: We're not going to tell you how to put this on your own calculator. You should be able to figure it out for yourself!

Problem 4: The program computes (a + b).

APPENDIX B

Testing and Troubleshooting

Mistakes are easy to make; even the most experienced programmer makes them. After you have written your program and keyed it into the calculator, it only makes sense to test it. If it doesn't seem to be working, you must then find your mistakes and correct them.

To test your program, try it sevral times with several different inputs. For most programs, there will be several inputs for which you already know (or can easily calculate by hand) the numbers that the program should output. If one of these inputs fails to produce the expected output, then you know something is amiss.

But say the program passes all your tests; is that a guarantee that everything is all right? Probably not. In most cases, the best you can do practically is to choose a small number of inputs to test, which together have a good chance of catching any error.

Here is our guideline: Make sure that each and every program step gets executed in at least one of your tests. Consider the program in Fig. B1, which accepts as input a number x. Here, A and B are blocks of program steps. If you tested only positive values of x, the block B of program steps could be full of mistakes, but you wouldn't realize it, because block B never gets used when the input x is positive.

Your more intricate programs will have many tests, branches, and subroutines. Each time you test an input, examine your flowchart to see which parts of it were tested. Try to find inputs that lead program execution to the yet untested portions of the flowchart, until *every* step is tested.

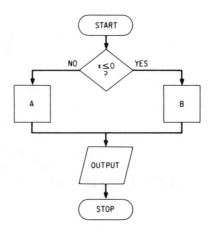

Fig. B1   Flowchart for troubleshooting

At times, you will have no idea what output to expect for any input, but ways still exist to test an input. For example, even though you cannot predict the output, there might be a simple check to determine if it is consistent or not. At the very least, there will always be some outputs that are obviously wrong. You will always have a general idea about the nature of the expected output. In some programs, for example, the output should always be a whole number; thus, you will know that something is wrong if your program outputs 31.52. In many programs, you will have a general idea as to the size of the output. If your program outputs a suprisingly small or surprisingly large number, you should get suspicious.

So much for testing your program. If it passes a large number of tests, it is *probably* okay. But what if it produces a definitely incorrect output, or even produces no output and just gives an "error" message? Then your program has at least one, and possibly many, mistakes in it. How do you find them?

The basic technique for troubleshooting a program is to execute it one step at a time, examining the display after each step to see if it has been executed correctly (see your manual on how to do this). You should have your flowchart and your program listing at hand while you are stepping through the program so that you can more easily keep track of where you are. As the calculator executes each step, use your flowchart and program listing to determine what it ought to do, and then compare this with the resulting display. When you arrive at a subroutine, most calculators will not let you single-step through it but will execute it all at once. If the subroutine appears to give the correct output, it's probably okay, and you can go on. If you suspect the subroutine is not working as it should, however, you should separately single-step through it. Many programmers test each subroutine all by itself as it is written, and you should definitely do likewise in programs using many complicated subroutines.

You will undoubtedly find it tedious to single-step through a program; we certainly do. In long programs, where single-stepping will take a great deal of time, you should try to make educated guesses as to which parts of the program might contain errors. Then you can carefully single-step through the suspect portions.

If you are at any point uncertain whether a particular portion of the program has been executed or not in the original test, you can insert an extra step or two in that part of the program to mark it. The easiest step to insert is setting an unused flag. Then run your program again, and see if the flag was set or not, to determine if the questionable portion of the program was executed. (If you have no flags, you can use an extra register; see Sec. 1.7.)

Now, when you find an error, consult your manual on how to remove the offending program steps and insert the correct ones. And when you've corrected all the mistakes you've found, you are not done yet; there could be more. Force yourself to go back and test your program again, paying careful attention to inputs that were previously troublesome.

Program Listings

In this Appendix, we give program listings for a few programs from Chap. 3 that fit very tightly onto a calculator with only 50 steps of program memory. All programs are written for the HP-25 (or HP-33E/C). Listings for the following programs are included: (1) blackjack dealer, (2) hockey, (3) jelly beans, (4) Wythoff's game, (5) nim, and (6) number jotto.

In blackjack dealer (Fig. C1), you should initialize 0 at 1.444444444, 1 at 16, 5 at 52, and 6 with seed. Each push of $\boxed{\text{R/S}}$ draws a card and returns a 0 when the deck is exhausted. The program has to be reinitialized (except for 6) when a new deck is used.

In hockey (Fig. C2), the program has been taken exactly from the flowchart, except for two things: (1) r_g is entered directly in the program in step 15, saving one storage register; and (2) steps 15 through 22 represent a tricky way of getting a 1 into m_7 if $|x_g| < r_g$ or a -1 if $|x_g| \geq r_g$. Note that $|x_g| \geq r_g$ if and only if $1 \geq r_g/|x_g|$.

In jelly beans (Fig. C3), the following steps are needed to initialize the program: Store 10 in r_3, store the smallest number of A in r_5, store the largest one in r_6, and store the set A as a decimal of 0's and 1's in r_7, as described in the text. To learn the calculator's move, enter the number of jelly beans left. It will display the number of jelly beans it "eats" on its turn.

In Wythoff's game (Fig. C4), after you have keyed in the program and reset the program counter, initialize by storing $(1/2)(\sqrt{5} + 1)$ in r_0. To learn the calculator's move from position (c,d), punch c, E↑, d, R/S. The calculator will output the resulting position (c',d'), after its move, in the x and y registers.

In nim (Fig. C5), key in the program and reset the program counter. To make the calculator move, store 0 in r_4; store 1 in r_7; store

01	RCL 6		23	0			*Memories*
02	π		24	×		0	cards < 10
03	+		25	FRAC		1	10 cards
04	5		26	STO 7			
05	y^x		27	LAST x		2	cards counted so far
06	FRAC		28	INT			
07	STO 6		29	STO + 2		3	card working
08	RCL 5		30	RCL 4		4	card drawn
09	×		31	RCL 2			
10	1		32	x ⩾ y		5	# of cards left in deck
11	STO 3		33	GTO 40			
12	STO −5		34	1		6	random # seed
13	+		35	STO + 3		7	unexamined cards
14	INT		36	RCL 7			
15	STO 4		37	x = 0			
16	RCL 1		38	GTO 00			
17	x ⩾ y		39	GTO 22		45	GTO 00
18	GTO 46		40	RCL 3		46	1
19	STO 2		41	CHS		47	STO − 1
20	RCL 0		42	10^x		48	1
21	FRAC		43	STO − 0		49	0
22	1		44	RCL 3		(50	GTO 00)

Fig. C1   Blackjack dealer

the position, encoded as a decimal .w (see text), r_6; and punch [R/S].
If the calculator outputs 0, it is in a losing position and has resigned.
Otherwise, to learn its move, store −1 in r_7, again store the position
.w in r_6, and punch [R/S]. When the calculator stops, its move is in the
x and y registers, encoded as follows:

"Remove all but x of the objects from a pile containing y objects"

To play number jotto (Fig. C6) with nine or less holes and nine
or less colors, the codemaker hides his code word in the calculator, and
it will give the code-breaker the number of black and white pegs for
each of his guesses. After keying in the program, fix 0, initialize the
program, and clear the register. Let h = the number of holes and c =
the number of colors. Label the colors with the e digits 1, 2, . . . , c.

START Punch
     [h], [STO], [0], [c], [CHS], [STO], [1]

CODEMAKER To hide word a,b, . . . ,s, punch
     [.], [a], [b], . . ., [s], [STO], [7], [CLx]

CODEBREAKER To guess word a,b, . . . ,s, punch
     [CLx], [STO], [2], [1], [STO], [3], [STO], [4]; then
     [.], [a], [b], . . ., [s], [CHS], [STO], [6], [R/S]

BLACK PEGS To see the number of black pegs, punch

$\boxed{\text{h}}$, $\boxed{\text{STO}}$, $\boxed{0}$, $\boxed{\text{RCL}}$, $\boxed{2}$

WHITE PEGS To see the number of white pegs, punch

$\boxed{\text{RCL}}$, $\boxed{3}$, $\boxed{+}$, $\boxed{-}$

To have the calculator choose a random code word, key in this program:

00				×		09			61
01	24	02	RCL 2	INT		10		14	01
02	15	73	π	1		11			01
03		51	+	+		12			51
04		05	5	STO + 7		13	23	51	07
05	14	03	y^x	1		14			01
06	15	01	FRAC	0		15			00
07	23	02	STO 2	STO ÷ 7		16	23	71	07
08	24	03	RCL 3	GTO 00		17		13	00

Store 0 in R_7, c in R_3, and an arbitrary number α, where $0 \le \alpha < 1$, in R_2. Punch R/S an h number of times. Key in number jotto program beginning again, initialize program, and guess.

(00	R/S)	26	RCL 5
01	STO 1	27	−
02	STO 5	28	ABS
03	R/S	29	RCL 6
04	STO 2	30	x < y
05	5	31	GTO 34
06	÷	32	0
07	STO 3	33	STO 7
08	STO 4	34	FIX 2
09	R/S	35	RCL 2
10	tan	36	PAUSE
11	STO × 3	37	RCL 1
12	RCL 2	38	PAUSE
13	×	39	RCL 3
14	STO −5	40	STO − 1
15	3 (r_g)	41	RCL 4
16	RCL 5	42	STO − 2
17	ABS	43	1
18	÷	44	STO − 0
19	1	45	RCL 0
20	x ≥ y	46	x ≠ 0
21	CHS	47	GTO 35
22	STO 7	48	FIX 0
23	6	49	RCL 7
24	STO 0	(50	GTO 00)
25	R/S		

Memories

0	loop counter
1	x-coordinate of puck
2	y-coordinate of puck
3	Δx
4	Δy
5	x_g
6	r_b
7	result of shot

Initial State of Memories

0	
1	
2	
3	
4	
5	
6	r_b
7	

Fig. C2 Hockey

(00	R/S)	23	1
01	STO 0	24	STO + 1
02	3	25	1
03	1/x	26	STO + 1
04	STO 1	27	STO − 0
05	RCL 1	28	RCL 0
06	STO 2	29	x < 0
07	RCL 7	30	GTO 34
08	STO −2	31	RCL 3
09	RCL 6	32	STO ÷ 1
10	STO 4	33	GTO 05
11	1	34	RCL 1
12	STO − 4	35	INT
13	RCL 4	36	1
14	x < 0	37	x = y
15	GTO 25	38	GTO 43
16	RCL 3	39	RCL 6
17	STO × 2	40	RCL 4
18	RCL 2	41	−
19	INT	42	GTO 00
20	STO − 2	43	RCL 5
21	x ≠ 2	44	GTO 00
22	GTO 11		

Fig. C3   Jelly beans

(00	R/S)	25	INT
01	x < y	26	RCL 5
02	x > < y	27	x = y
03	STO 2	28	GTO 39
04	x > < y	29	RCL 1
05	STO 1	30	RCL 0
06	−	31	×
07	STO 3	32	INT
08	RCL 0	33	LAST x
09	×	34	x = y
10	INT	35	GTO 39
11	STO 4	36	x > < y
12	RCL 1	37	1
13	x = y	38	+
14	GTO 41	39	RCL 1
15	x ≥ y	40	GTO 00
16	GTO 45	41	RCL 2
17	RCL 1	42	1
18	RCL 0	43	−
19	÷	44	GTO 00
20	INT	45	RCL 3
21	STO 5	46	RCL 4
22	−	47	+
23	RCL 0	48	LAST x
24	×	49	GTO 00

Fig. C4   Whythoff's game

(00	R/S)	17	1/x	34	GTO 21
01*	ε	18	STO 0	35	RCL 0
02	10^x	19	CLx	36	STO + 5
03	STO × 6	20	STO 5	37	GTO 21
04	RCL 6	21	2	38	RCL 7
05	x = 0	22	STO × 0	39	x < 0
06	GTO 49	23	STO ÷ 1	40	GTO 44
07	INT	24	STO ÷ 2	41	RCL 5
08	STO 3	25	RCL 1	42	STO 4
09	STO − 6	26	x = 0	43	GTO 01
10	RCL 4	27	GTO 38	44	RCL 3
11	x < y	28	FRAC	45	RCL 5
12	x < > y	29	STO − 1	46	x ≥ y
13	STO 1	30	RCL 2	47	GTO 01
14	x > < y	31	FRAC	48	GTO 00
15	STO 2	32	STO − 2	49	RCL 4
16	2	33	x = y		

* The "ε" in step 01 is either a 1 or a 2, as described in the text. If ε = 1, there are at most nine objects per pile. If ε = 2, there are at most five piles, with at most 99 objects per pile.

Fig. C5 Nim

Step Number ↓	Machine Code	Step Name ↓	Step Name ↓	Step Number ↓	Machine Code
00			GTO 37	25	13 87
01	24 07	RCL 7	10^x	26	15 08
02	23 41 05	STO − 5	STO + 4	27	23 51 04
03	01	1	↓	28	22
04	32	CHS	10^x	29	15 08
05	23 61 01	STO × 1	STO + 3	30	23 51 03
06	01	1	GTO 06	31	13 06
07	23 41 00	STO − 0	↓	32	22
08	15 08	10^x	−	33	41
09	23 61 05	STO × 5	x ≥ 0	34	15 51
10	23 61 06	STO × 6	STO + 3	35	23 51 03
11	24 00	RCL 0	GTO 06	36	13 06
12	15 41	x < 0	1	37	01
13	13 40	GTO 40	STO + 2	38	23 51 02
14	24 05	RCL 5	GTO 06	39	13 06
15	14 01	INT	RCL 1	40	24 01
16	23 41 05	STO − 5	x < 0	41	15 41
17	24 06	RCL 6	GTO 00	42	13 00
18	14 01	INT	STO 0	43	23 00
19	23 41 06	STO − 6	RCL 3	44	24 03
20	24 01	RCL 1	STO −3	45	23 41 03
21	15 41	x < 0	STO 5	46	23 05
22	13 32	GTO 32	RCL 4	47	24 04
23	22	↓	STO 6	48	23 06
24	14 71	x = y	GTO 03	49	13 03

Fig. C6 Number jotto

APPENDIX D

Notation

Frequently throughout the book we refer to particular keys on the calculator, using the standard HP or TI notations. Sometimes when these differ, we don't mention the alternative. Here is a list of equivalents for these keys:

HP	TI	
ABS	= |x|	Absolute value
CHS	= +/−	Change sign
D.MS	= H.MS	"Degrees, minutes, seconds"
E EX	= EE	Enter exponent
e^x	= INV ln x	Exponential function
FRAC	= INV INT	Fractional part
GSB	= SBR	Execute subroutine
RTN	= INV SBR	Return
STO +	= SUM	Register addition
STO −	= INV SUM	Register subtraction
STO X	= Prd	Register multiplication
STO ÷	= INV Prd	Register division
10^x	= INV log x	Inverse common log

Testing that involves two numbers uses the y register on HP's and the t-register on TI's. We occasionally use y interchangeably with t in the text.

In flowcharts and elsewhere, x refers to the contents of the display register; m_1, m_2, r_1, r_2, etc., refer to memory registers; m_1 or r_1 usually designates memory 1 (or 01); m_2 or r_2 usually designates memory 2, etc.

The notation $x \leftarrow R$ and (m_0) means "generate a random number between 0 and 1 from the (previous random) number in m_0 and place it in m_0 and the display." See Sec. 1.11 for details.

The notation "←" appearing in the flowcharts means "place the number indicated on the right in the memory indicated on the left." See Appendix A for further explanation.

Glossary

Address (*n.*) The name of a memory. An *addressable memory* is one whose contents can be recovered or manipulated by reference to its name.

Algebraic logic (*n.*) The method of describing a sequence of algebraic operations by the use of parentheses and a hierarchy of operations. Most calculators use either some form of algebraic logic or RPN.

Algorithm 1. (*n.*) A computational recipe; in particular, a recipe that can be programmed onto a computer and that stops in finite time for each legal input. 2. (*n.*) A procedure that terminates.

Branch 1. (*n.*) A place in a program where control can go one of two ways, depending on the outcome of a test. 2. (*v.*) To send control to another part of the program, usually as the result of a test.

Call (*v.*) To order execution, usually of a subroutine.

Clear 1. (*n.*) The lowered state of a flag. 2. (*v.*) To lower (a flag). 3. (*v.*) To erase; "to clear a memory."

Control 1. (*n.*) The operator in control of the calculator, either the program or the user. 2. (*n.*) That part of the program being executed at a given moment.

Counter (*n.*) A memory used to count the number of times that some part of a program has been executed (frequently used to control loops).

Counterexample (*n.*) A specific input that fails to produce the conjectured or expected output.

Cue (*n.*) A signal displayed by the calculator during a temporary halt in program execution, usually used to remind the user what additional data is required.

Data (*n.*) Information; specifically, numerical information processed by the calculator. *Input data* is information supplied by the user to the calculator. *Output data* is information supplied to the user by the calculator.

313

Decrement   1. (*n.*) A fixed amount by which another quantity is repeatedly decreased 2. (*v.*) To decrease by a decrement. (See also
Increment.)

Dependent variable   (*n.*) See *Independent variable*

Display   (*n.*) That part at the top of your calculator where all those
numbers magically show up.

Dividend   See *Division*

Division   (*n.*) The act of dividing a number y (called the *dividend*) by
a number x, where x $\neq$ 0 (called the *divisor*). There two types: (1)
Real division, which outputs the *quotient,* y/x, and (2) *quotient and
remainder division,* used only when x and y are integers. It outputs an integer q, called the *quotient,* and an integer r, called the
remainder, so that y = xq + r and 0 $\leqslant$ r < |x|. (This operation does
not appear on most calculators and must be programmed in; see,
for example, Sec. 2.11.)

Divisor   See *Division*

Evaluate.   (*v.*) To compute the value of a dependent variable from a
given value(s) of the independent variable(s).

Exponential function   (*n.*) The function, x $\leftarrow$ e^x, or, more generally,
any function of the form, x $\leftarrow$ a^x, where a is a constant; for example, x $\leftarrow$ 10^x. Exponential functions are the inverses of log
functions.

Feedback   (*n.*) Output that is fed back into the program, usually for
the purpose of refining a computation.

Fixed point   (*n.*) The method of representing each number by a string
of one or more digits, followed by a decimal point, followed by a
string of digits (compare with *Floating point*).

Flag   (*n.*) A memory with only two possible states ("set" and
"clear") that is used (1) in conjunction with flag-testing functions
to cause conditional branches, and (2) diagnostically, to check the
flow of control in a program or to check the state of the calculator.

Floating point (scientific notation)   (*n.*) The method of representing
each nonzero number in the form, $\pm(\alpha \times 10^n)$, where $1 \leqslant \alpha < 10$
and n is an integer (compare with *Fixed point*).

Function   (*n.*) A process that unambiguously assigns to certain numbers, or sets of numbers, other numbers, or sets of numbers. Many
functions such as the function that assigns to every nonnegative
number its square root can be executed on your calculator by a
few key strokes; many others can be programmed into your calculator.

Increment   1. (*n.*) A fixed amount, frequently 1, by which another quantity is repeatedly increased; a negative increment is called a *decrement*. 2. (*v.*) To increase by an increment.

Independent variable (*n.*) A variable in a formula (or formulas) that may take any of a range of values and whose value, perhaps in conjunction with the values of other independent variables in the formula, in turn determines the value of another variable (or variables), called a *dependent* variable. Independent variables can be thought of as the input of a formula and dependent variables as the output.

Indirect addressing (*n.*) Addressing a memory through the use of an intermediate memory (see *Indirect addressing* for details).

Initial state (*n.*) The state the calculator must be in at the start of program execution; in particular, the state of the memories.

Initialize (*v.*) To place the calculator in its initial state.

Integer (*n.*) A whole number, positive, negative, or zero.

Input (*n.*) The data to be fed to the calculator at the beginning, and sometimes during, program execution.

Iterate 1. (*v.*) To repeat an algorithm or other computational procedure with the output from the last run as the input for the next run; more generally, simply to repeat. 2. (*n.*) The output of an iteration.

Iteration (*n.*) A single repetition of an iterated procedure.

Key 1. (*n.*) Any one of the buttons on the face of a calculator. 2. (*v.*) To perform a sequence of keystrokes.

Keystroke (*n.*) The act of depressing and releasing a single key on the calculator.

Label (*n.*) A program step whose only purpose is to assign a name to its location in program memory. Execution of a label has no affect on the state of the calculator, but other parts of the program can transfer execution to the label and hence to the steps following it in the program.

Logic (*n.*) The form in which a calculator encodes, or requires the user to encode, algebraic operations. There are two standard forms, *algebraic logic* (see *Algebraic logic*) and *RPN* (see *RPN*).

Loop (*n.*) A sequence of program steps that is repeatedly executed, usually until some test sends execution "out of the loop."

Lower bound (*n.*) A number that is known to be smaller than all possible values of some variable or unknown.

Mantissa 1. (*n.*) A number between 1 and 10 used as a component of scientific notation, for example, the mantissa of 3.68×10^{12} is 3.68; see also *Floating point notation*. 2. (*n.*) (archaic) A component of precalculator logarithms. The log of 3.68 × 10^{12} is 12.5658 and .5658 is the mantissa. In the past, tables for logarithms listed only mantissas.

Memory (also called *Register*)   (*n.*) A location in the calculator in which information can be stored and from which information can be recalled. There are many types: *data register,* which stores numbers in floating point format; *program memory,* which stores program steps; *addressable memory* (see *Address*); *ROM* (*R*ead *O*nly *M*emory), a memory from which information may be recovered but whose contents cannot be altered.

Memory arithmetic   (*n.*) See *Register arithmetic.*

Model   1. (*n.*) A physical or mathematical picture/encoding/replica used for experimental purposes; in particular, an encoding which can be placed on a computer or calculator. 2. (*v.*) To produce a model. 3. (*v.*) To operate a model.

Multiple storage   (*n.*) Storing more than one number in a memory (see Sec. 1.14).

Operator   1. (*n.*) A function. 2. (*n.*) A function requiring more than one number as input, for example, addition. 3. (*n.*) A function whose input is another function, for example, a routine that finds the roots of a function.

Output   (*n.*) The data a calculator presents to the user at the end, and sometimes during, program execution.

Overflow   (*v.*) To perform an operation whose result is a number of absolute value so large that it will not fit into a data register.

Π-*notation* (pi notation)   (*n.*) A notation for long, systematically generated products (see Sec. 1.8).

Parameter   (*n.*) An independent variable, or a variable whose value must be known in order to compute the value of some other variable; more generally, a piece of data that affects other data.

Parametric equations   (*n.*) A collection of equations defining a collection of dependent variables in terms of one or more independent variables called the *parameters,* for example, x = r cosθ and y = r sinθ form a pair of parametric equations defining the x and y coordinates of points on a circle of radius r in terms of the parameter θ.

Procedure   (*n.*) A precisely defined sequence of instructions.

Program   1. (*n.*) A sequence of keystrokes to be stored in the program memory for the calculator to execute sequentially; they are usually expected to have some definite purpose. 2. (*v.*) To store a program into the program memory.

Program memory   See *Memory.*

Program pointer (*n.*) A register that keeps track of which step in a program is to be executed next.

Programmer (*n.*) The person who writes a program.

Quotient See *Division*.

RPN (*R*everse *P*olish *N*otation) (*n.*) A language for expressing algebraic operations unambiguously without parentheses. Most calculators use either RPN or the more traditional algebraic logic.

Recursion (*n.*) A rule for calculating each number in a sequence of numbers from the previous numbers in the sequence.

Register (*n.*) A memory. Registers are of several types: *storage registers, program memory registers,* the *x-register,* etc. See also *Memory*.

Register arithmetic (*n.*) Arithmetic performed on the contents of a memory without calling it into the display. STO +, STO × (HP), SUM and INV SUM (TI) are some register arithmetic operations.

Remainder See *Division*.

Return 1. (*v.*) To return from a subroutine. 2. (*v.*) To produce as output ("This subroutine returns the sum of the values in m_1 and m_2").

Σ-*notation* (sigma notation) (*n.*) A notation for long, systematically generated sums (see Sec. 1.8).

Scientific notation See *Floating point*.

Seed (*n.*) A number used to initialize a random number generator or other iterative procedure.

Sequence (*n.*) A list of numbers; a sequence may be infinite.

Set 1. (*n.*) The raised state of a flag. 2. (*v.*) To raise (a flag).

Simulate (*v.*) To model on a computer or calculator.

Step 1. (*n.*) A single line of a program. 2. (*v.*) To execute a single line of a program.

String (*n.*) A list of symbols, usually digits.

Subroutine (*n.*) A mini-program that can be used by the main program several times. Most calculators will automatically transfer execution back to the proper place in the main program after the execution of a subroutine (see Sec. 1.7).

Test 1. (*n.*) A comparison of two numbers (or of a single number with some fixed number, usually 0) in a program (see also *Test function*). 2. (*v.*) To read the state of a flag.

Test function (*n.*) A program step that executes a test and prepares to branch (see *Test* and *Branch*).

Underflow   (*v.*) To perform an operation whose result is a nonzero number of absolute value so small that it cannot be distinguished from 0.

Unit   (*n.*) A fixed quantity or measure used as a standard. Conventional units include feet, kilograms, watts, etc. Often a programmer will introduce his own convenient units for a particular problem.

Upper Bound   (*n.*) A number that is known to be larger than all possible values of some variable or unknown.

User   (*n.*) A person who uses a program.

User-definable keys   (*n.*) A key whose function can be defined by the user, rather than the manufacturer, of the calculator.

Value   (*n.*) A particular number substituted for a variable. Independent variables can be *assigned* values. Dependent variables *take on* values when the values of the associated independent variables have been assigned.

Variable   1. (*n.*) A symbol that may take any of a collection of numerical values. 2. (*n.*) A memory that may store any of a collection of numbers in a program.

x-register   (*n.*) The memory that stores the number appearing in the display.

Index

*(Entries marked with an * are defined in the glossary.)*